98-HHE-2

FATE AND PERSISTENCE OF PATHOGENS SUBJECTED TO ULTRAVIOLET LIGHT AND CHLORINE DISINFECTION

by:

Karl G. Linden, Ph.D.
Department of Civil and Environmental Engineering
Duke University

James D. Oliver, Ph.D.
Department of Biology
University of North Carolina, Charlotte

Mark D. Sobsey, Ph.D., and Gwy-Am Shin, Ph.D.
Department of Environmental Science and Engineering
University of North Carolina, Chapel Hill

2004

WERF

The Water Environment Research Foundation, a not-for-profit organization, funds and manages water quality research for its subscribers through a diverse public-private partnership between municipal utilities, corporations, academia, industry, and the federal government. WERF subscribers include municipal and regional water and wastewater utilities, industrial corporations, environmental engineering firms, and others who share a commitment to cost-effective water quality solutions. WERF is dedicated to advancing science and technology addressing water quality issues as they impact water resources, the atmosphere, the lands, and quality of life.

For more information, contact:
Water Environment Research Foundation
635 Slaters Lane, Suite 300
Alexandria, VA 22314-1177
Tel: (703) 684-2470
Fax: (703) 299-0742
www.werf.org
werf@werf.org

This report was co-published by the following organizations. For non-subscriber sales information, contact:

Water Environment Federation
601 Wythe Street
Alexandria, VA 22314-1994
Tel: (800) 666-0206
Tel: (703) 684-2452
Fax: (703) 684-2492
www.wef.org
pubs@wef.org

IWA Publishing
Alliance House, 12 Caxton Street
London SW1H 0QS, United Kingdom
Tel: +44 (0) 20 7654 5500
Fax: +44 (0) 20 7654 5555
www.iwapublishing.com
publications@iwap.co.uk

Library of Congress Catalog Card Number: 2003108358

IWAP ISBN: 1-84339-699-8 WEF ISBN: 1-57278-182-3

Duke University, University of North Carolina

ACKNOWLEDGMENTS

The authors would like to acknowledge the assistance of students and technicians who assisted during the research project. From Duke University: Lance Batch, master's student; Sharleen Johnson, master's candidate; and Zuzana Bohrerova, post-doctoral research associate. All contributed to parts of this study. From the University of North Carolina–Charlotte: Maya Dagher and Lee Anne Wilber performed the laboratory work as part of their degree programs. From the University of North Carolina–Chapel Hill: Douglas Wait, research analyst and laboratory manager; Gregory Lovelace, research associate; Dorothy Thompson, research assistant; and Angela Couliette, graduate research assistant. All contributed to the experimental work.

The project team would also like to thank Mike Arrowood, Division of Parasitic Diseases, Centers for Disease Control and Prevention [Atlanta, Georgia] for the generous gift of fluorochrome-labeled C3-C3 antibody; and Dr. W.O.K. Grabow, University of Pretoria [Pretoris, South Africa], a member of the Podoviridae family, for the generous gift of coliphage V1.

Finally, the authors would like to thank the Project Subcommittee for their insight and for suggestions made throughout the research project and the for the help of the WERF project managers, Jean Tribull and Jami Montgomery.

Report Preparation

Principal Investigators:
Karl G. Linden, Ph.D.
Duke University

James D. Oliver, Ph.D.
University of North Carolina–Charlotte

Mark D. Sobsey, Ph.D.
University of North Carolina–Chapel Hill

Project Team:
Gwy-Am Shin Ph.D.
University of North Carolina–Chapel Hill

Banu Ormeci Ph.D.
Duke University

Project Subcommittee

Jennifer L. Clancy, Ph.D., Chair
Clancy Environmental Consultants, Inc.

Charles Gerba, Ph.D.
Microbiology Program, University of Arizona

Walter Jakubowski, Ph.D.
WaltJay Consulting

Larry Loomis
David Trudil, Ph.D.
New Horizon Diagnostics Corporation

William A. Yanko

Water Environment Research Foundation Staff

Director of Research: Daniel M. Woltering, Ph.D.
Project Manager: Jami Montgomery

ABSTRACT AND BENEFITS

Abstract:

Disinfection of wastewater protects the public from potential exposure to pathogenic microorganisms that would otherwise be present in wastewater effluent that is discharged into water bodies that may be used for recreation or a source of drinking water. Chlorine and ultraviolet (UV) light are two common forms of disinfection. However, microorganisms differ in their susceptibility to UV and chlorine disinfectants. Further, disinfection of microorganisms in wastewater presents challenges that are inherent to the water matrix, such as pathogens associated with particles. Water quality may also impact disinfection effectiveness. Understanding how different classes of pathogens respond to UV and chlorine disinfection processes in wastewater treatment can help with development of strategies for optimizing wastewater treatment.

This study investigated bacteria, viruses, and protozoan pathogens. All species of bacteria tested were susceptible to both UV and chlorine, despite differences in antibiotic resistance and tendency to aggregate. Upon exposure to disinfection conditions that could indicate viability of the bacteria tested, but not culturability using common methods, UV and chlorine were effective in eliminating the capability of viable but nonculturable bacteria to resuscitate and become re-infective. *Clostridium* spores were resistant to free chlorine (FC) and UV disinfection but were susceptible to long exposure to monochloramine. *Cryptosporidium* was resistant to all chlorine forms but very susceptible to UV irradiation. Pathogenic and indicator viruses tested were very susceptible to FC and UV disinfection. UV radiation throughout the 200 nm to 300 nm range was effective for inactivation of viruses and *Cryptosporidium parvum*. However, wavelengths between 260 nm and 270 nm and <220 nm appeared to be more effective for viruses, suggesting a possible advantage for polychromatic UV sources. Sequential disinfection strategies were proposed and tested to enhance inactivation of various microorganisms. One scenario integrated UV disinfection followed by dynamic chloramination through the addition of FC and subsequent transformation to combined chlorine. UV and chlorine were both effective for disinfection of coliform in wastewater, but chlorine was found to be more effective during long contact times for inactivation of particle-associated coliform. In addition to coliform, both *C. parvum* and *Salmonella typhimurium* were identified as particle associated in wastewater using molecular approaches developed to detect microbes in environmental samples.

Benefits:

- Improves understanding of the mechanisms of disinfectant action and disinfection resistance for important pathogens and indicators in the water environment.
- Recommends disinfection process control that will enhance disinfectants currently in use so that they can effectively inactivate persistent pathogens.
- Assesses the limits of the methodologies currently used for enumerating organisms.
- Develops and refines methodologies to detect the presence of pathogens embedded in wastewater particulate matter, including molecular-based technologies.

Keywords: ultraviolet irradiation, chlorination, dynamic chloramination, pathogens, reactivation, particles, wastewater, FISH

TABLE OF CONTENTS

LIST OF TABLES

LIST OF FIGURES

LIST OF ACRONYMS

ANOVA	analysis of variance
AODC	acridine orange direct count
ATP	adenine tri-phosphate
AV2	adenovirus type 2
BSA	bovine serum albumin
CB	collimated beam
Cfu	colony-forming unit(s)
cm	centimeter(s)
CT	concentration * time
CTC	hydrolysis of 5-cyano-2,3-di-4-tolyl-tetrazolium chloride
CVB4	Coxsackievirus B4
CY3	Cyanine 3
DAL	double agar layer
DAPI	4',6-diamidino-2-phenylindole, dihydrochloride
DI	deionized water
DIN	Deutsche Industrialen Normalen
DMEM	Dulbecco's Modified Eagles Medium
DNA	deoxyribonucleic acid
DPD	N,N-diethyl-p-phenylenediamine
DVC	direct viable count
EDTA	ethylenediaminetetraacetic acid
EGTA	ethylenebis(oxyethylenenitrilo)tetraacetic acid
FC	free chlorine
FISH	fluorescent in-situ hybridization
FITC	fluorescein isothiocyanate
HAV	hepatitis A virus
HCT-8	human colon tumor cells
HOCL	hypochlorous acid
IES	Illumination Engineering Society
IT	intensity * time
J	joule(s)
L	liter(s)
LB	Luria Bertani
LP	low pressure
μW	microwatt(s)
MDCK	Madin-Darby Canine Kidney
MF	membrane filtration
mg/L	milligram(s) per liter
min	minutes
mJ	millijoule(s)
mL	milliliter(s)
mm	millimiter(s)
mM	millimolar

MP	medium pressure
MPN	most probable number
MS2	MS2 bacteriophage
mW	milliwatt(s)
N_d	final concentration of microorganisms
N_o	initial concentration of microorganisms
nm	nanometer(s)
NPAC	non-particle associated coliform
PAC	particle-associated coliform
PBS	phosphate-buffer solution
PCR	Polymerase Chain Reaction
PFU	plaque-forming unit(s)
PV1	Poliovirus 1
pH	negative logarithm of the effective hydrogen-ion concentration
RAPD	randomly amplified polymorphic
RNA	ribonucleic acid
ROS	reactive oxygen species
rpm	rotations per minute
SAL	single agar layer
SDS	sodium dodecylsulphate
TMR	tetramethyl rhodamine
TOC	total organic carbon
TSA	tryptic soy agar
TSB	tryptic soy broth
UNC	University of North Carolina
U.S. EPA	United States Environmental Protection Agency
UV	ultraviolet
VBNC	viable but nonculturable
W	watt(s)
WRF	water reclamation facility
WWTP	wastewater treatment plant

EXECUTIVE SUMMARY

Chlorination and ultraviolet (UV) irradiation are the two most prominent methods of wastewater disinfection in the United States. Chlorination in wastewater is usually accomplished using either chlorine gas or sodium hypochlorite, which result in forms of free and/or combined chlorine available for disinfection. UV irradiation in wastewater uses either low- or medium-pressure UV mercury vapor lamps, which generate germicidal wavelengths including 254 nm. Both methods of disinfection have proven effective in meeting federal wastewater effluent standards based on the reduction of indicator organisms in the finished effluents. However, the fate of specific enteric pathogenic bacteria, viruses, and protozoans in wastewater disinfection processes and the extent to which they elude disinfection or detection in wastewater are not well understood or documented.

Many factors relate to the fate of pathogens in disinfection processes, but the two of the major factors are likely 1) morphological characteristics of the specific organism that protect it from receiving lethal disinfectant doses, and 2) water quality characteristics that prevent adequate disinfectant persistence or its contact with and penetration of the pathogen of concern. Another concern is that the ability to detect pathogens surviving disinfection using currently accepted microbial culture techniques may be compromised when attempting to enumerate organisms that were either non-lethally injured and may later recover or repair, or induced into the viable but nonculturable (VBNC) state. The false negatives (assuming a potentially viable pathogen has been inactivated because it cannot be cultured) resulting from nondetection and subsequent repair or induction into the VBNC response compromise the ability to predict disinfection performance and protect public health.

Given these challenges, it is necessary to optimize the management of existing facilities, investigate new and emerging disinfection technologies (e.g. UV irradiation), and develop detection methods for VBNC microbes in order to make wastewater disinfection processes more effective against all waterborne pathogens. This will better protect the public from possible health risks. Along with these efforts, the germicidal efficacy and modes of action of these new and emerging UV technologies should be characterized to optimize their use in actual wastewater disinfection practice.

This research studied the fate and persistence of pathogens subject to disinfection by examining the potential mechanisms of pathogen inactivation and resistance, the occlusion of pathogens and their protection from disinfectants, and the inability to detect total viable pathogen presence following disinfection. The research also investigated ways to optimize the use of current and emerging disinfection technologies to improve disinfection performance in wastewater.

Results indicate that both low- and medium-pressure UV irradiation are very effective for inactivation of *Cryptosporidium parvum* oocysts. In fact, >3 $\log_{10}$ inactivation was achieved in wastewater with a UV dose of only 3 mJ/cm^2. Both UV technologies were also shown to be effective for inactivation of various enteric viruses. However, the viruses were generally more resistant to UV inactivation (often more than 10 times more resistant) than enteric bacteria and protozoans. The use of polychromatic UV irradiation from medium-pressure UV lamps may have an advantage over the use of monochromatic (UV 254 nm) low-pressure UV lamps for inactivation of enteric viruses because wavelengths in the 260 nm to 270 nm range and <220 nm appeared to be more effective than UV wavelengths of 254 nm for inactivation of a model MS2 virus. Inactivation of *Cryptosporidium* was most effective in the 250 nm to 270 nm range, which includes both the low- and medium-pressure output ranges. Even though many microorganisms exhibit the ability to repair UV-damaged DNA and restore infectivity, these results show that UV-inactivated *C. parvum* oocysts are not able to restore their infectivity in cell culture host cell lines following exposure to either light (photoreactivation) or dark DNA repair protocols.

These findings also indicate that inactivation of human enteric viruses and bacteriophages by UV irradiation are not reliably predicted by the type and size of the virus or its nucleic acid genome. Two human enteric viruses, poliovirus 1 and Coxsackievirus B4 that are small (~27 nm) and have RNA genomes of about 7.5 kb were relatively sensitive to low-pressure (LP) UV radiation (>4 $\log_{10}$ inactivation within a UV dose of 30 mJ/cm^2). However, an RNA bacteriophage with similar overall size but a smaller RNA genome (~4 kb), coliphage MS2, was relatively resistant to UV irradiation (~2 $\log_{10}$ inactivation with a UV dose of 30 mJ/cm^2). In contrast, the two large (50 nm to 100 nm) DNA bacteriophage with linear, double-stranded DNA were relatively sensitive to LP UV radiation (>3 $\log_{10}$ inactivation with a UV dose of 10 mJ/cm^2). However, a human pathogenic virus with similar size and linear double-stranded DNA, adenovirus 2, was highly resistant to LP UV radiation (~2 $\log_{10}$ inactivation with a UV dose of 60 mJ/cm^2). Results suggest that there is no strong association between virion size and genetic composition of enteric viruses and their response to LP UV irradiation. Similarly, experiments with isogenic pairs of bacteria that differed in resistance to antibiotics showed that these factors were not associated with increased survival or resistance to UV disinfection or chlorination.

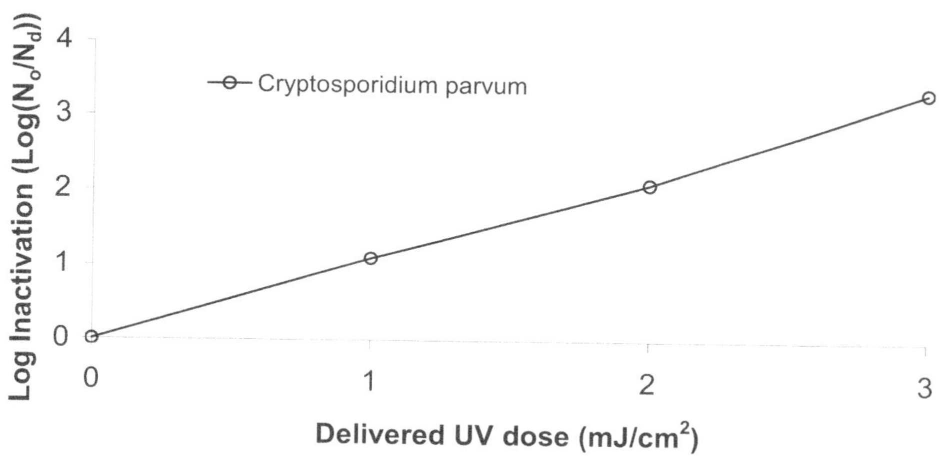

Figure ES-1. Inactivation of *C. parvum* Oocysts in Wastewater by LP UV Radiation.

Another focus of this research was to determine whether cells enter into the VBNC state upon irradiation with UV light or exposure to chlorine. In this state, cells are no longer culturable by normal microbiological methods (e.g., most probable number or plate counts), but can be demonstrated to be alive ("viable"). Cells are induced into this state by a variety of natural environmental factors, including low and high temperature, high osmotic conditions, and even sunlight. Some cells in this state have been demonstrated to "resuscitate" back to the actively metabolizing and culturable state, often by a simple reversal of the inducing factor. UV irradiation does appear to induce cells to become nonculturable, but the DNA crosslinking that occurs is likely to be substantial. No indication of recovery (resuscitation) from this state was evident. Chlorination also induces nonculturability. However, with dynamic chlorine/chloramine treatment in wastewater, no evidence of resuscitation was observed.

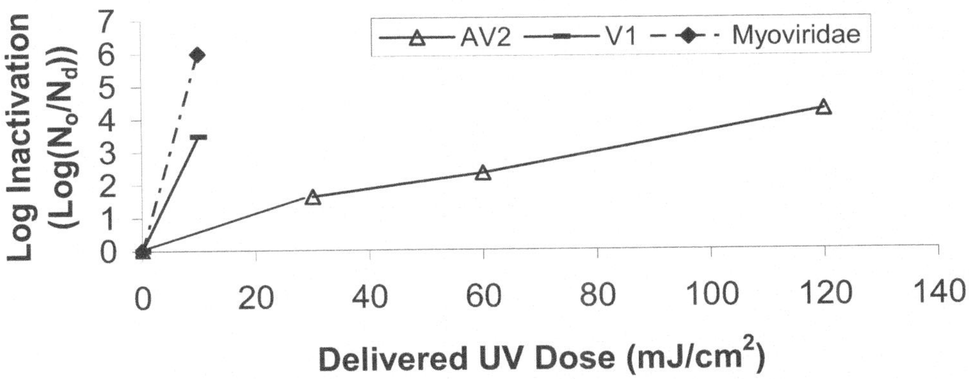

Figure ES-2. Inactivation of DNA Viruses by LP UV: AV2, bacteriophage V1, Myoviridae.

The issue of particle association of microbes affecting disinfection efficacy was tested under numerous scenarios. A protocol for detecting and extracting particle-associated coliform was developed as part of developing a better understanding of particle association with pathogens. It appears that aggregation and/or association shielded indigenous coliphage in wastewater and protected them from lethal exposure to UV irradiation, reducing the ultimate UV disinfection efficacy. Similar results were obtained for coliform in wastewater. Although a significant fraction of these microbes could be inactivated, reduction to the level of nondetection was not achievable even at relatively high doses of UV light.

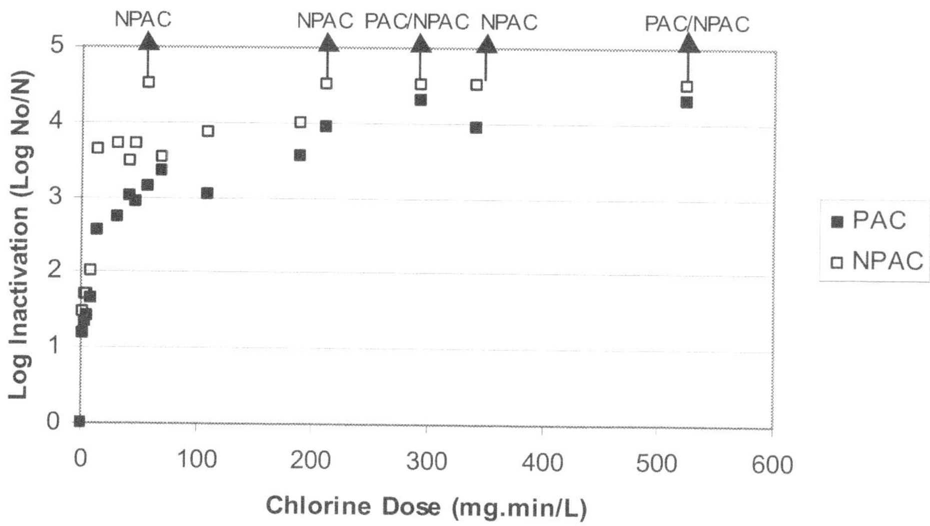

Figure ES-3. Dose-survival Relationship of Particle-associated Coliform (PAC) and Nonparticle-associated Coliform (NPAC) Under Chlorine Disinfection.

Because most pathogens are not present in numbers high enough to detect particle association using methods utilized for coliform, a molecular biological approach was used to identify pathogens in wastewater and visualize the degree of particle association. The method chosen for identification of pathogens associated with particles and flocs in wastewater was fluorescent in-situ hybridization (FISH). This method was modified to minimize the impact on particle/floc structure and allow visualization with minimal disturbance to the floc structure. It was visually shown that the pathogens *Salmonella* and *C. parvum* were present and associated within wastewater flocs, in a fashion similar to that found for coliform bacteria. Although disinfection testing of these particle-associated pathogens was not possible, it may be assumed that these microbes would also be protected from UV and chlorine disinfection. However, the extent of the significance of this particle association was not evaluated and would likely be source and process specific.

Use of multiple disinfectants in series was hypothesized to be an effective strategy for inactivation of the wide range of pathogens found in wastewater. Results indicate that a significant level of inactivation of somatic coliphage and *C. parvum* oocysts can be achieved by sequential application of free chlorine (FC) and LP UV. However, *C. perfringens* spores, a fecal indicator microbe, were shown to be extremely resistant to this sequential disinfection scheme. However, sequential application of LP UV and monochloramine showed a greater level of inactivation of *C. perfringens* spores as well as other microbes tested. Therefore, an approach that utilizes UV disinfection followed by FC dosing and subsequent formation of monochloramine (due to ammonia in the wastewater) along with long contact time should be capable of achieving significant inactivation of most microorganisms within a practical range of UV and free/monochloramine doses. Extended contact time with chlorine was also found to be effective in achieving inactivation of particle-associated coliform in wastewater. However, because protozoan pathogens such as *C. parvum* oocysts are highly resistant to chlorination, non-UV disinfected particle associated *C. parvum* oocysts may still be persistent in wastewater following chlorination only. Therefore, if complete inactivation of chlorine resistant microbes is required, use of UV-disinfection systems in wastewater treatment processes can be very effective but should be optimized or augmented (perhaps with filtration) to compensate for the reduction of disinfection efficacy due to aggregation and/or particle association.

These results can be directly applied to optimization of current wastewater disinfection practice with UV irradiation and chlorine. For example, a wastewater utility discharging their treated effluent to a sensitive water body (e.g., one used for drinking water intake) may want to employ a sequential UV-chlorine/chloramine disinfection scheme to ensure adequate inactivation of most pathogenic viruses and protozoans. As another example, a wastewater utility that is experiencing problems meeting effluent standards may need to investigate the particle-associated microbes and adjust their upstream processes to minimize formation of microbes associated with particles. Finally, although the use of FC/chloramines sequential disinfection in a dynamic chloramination mode by a wastewater treatment plant could enhance the overall removal of bacteria, viruses, and spores with extended contact times, this scenario is not capable of disinfecting *C. parvum* oocysts. Therefore, for some wastewater utilities that are particularly susceptible to *Cryptosporidium*, the addition of UV disinfection process would be a necessary step to protect downstream users from the risk of *Cryptosporidium* exposure.

CHAPTER 1.0

EXPERIMENTAL PLAN AND APPROACH

1.1 Overview

Chlorination and ultraviolet (UV) irradiation are the two most commonly used disinfectants in wastewater treatment in the United States. Current technology for UV irradiation employed in wastewater includes low pressure (LP) and medium pressure (MP) UV mercury vapor lamps. Disinfection of wastewater with chlorine is commonly accomplished using either chlorine gas or sodium hypochlorite, resulting in free and/or combined chlorine available for disinfection. Both methods of disinfection have proven effective in meeting effluent standards based on survival of indicator organisms. However, the fate of specific enteropathogenic bacteria, human enteric viruses, and enteric protozoans and the extent to which they elude disinfection or detection in wastewater is not well understood or documented.

The ways pathogens persist following disinfection can be related to 1) morphological characteristics of the specific organism that protect it from receiving lethal disinfectant doses, or 2) water quality characteristics that prevent adequate penetration of a disinfectant to the pathogen. Furthermore, the ability to detect pathogens surviving disinfection using accepted microbial culturing techniques may be compromised when attempting to enumerate organisms that were either nonlethally injured and may later recover or repair, or induced into the viable but nonculturable (VBNC) state. The false negatives (assuming a potentially viable pathogen has been inactivated because it cannot be cultured) resulting from non-detection and subsequent repair or induction into the VBNC response compromise the ability to predict disinfection performance and protect public health. Optimized management of existing facilities and investigation of UV irradiation and chlorination disinfection technologies are needed to assess their effectiveness against all waterborne pathogens. The germicidal efficacy and modes of action of UV irradiation and chlorination and other emerging technologies must be characterized in order to optimize their use in wastewater treatment. Promising disinfection systems to explore are: 1) high-intensity UV radiation from MP mercury lamps, and 2) controlled chlorination with free chlorine (FC) converting over time to chloramines.

This research approaches the study of fate and persistence of pathogens subject to disinfection through examination of the specific mechanisms of pathogen inactivation and resistance, the occlusion of pathogens from disinfectants, and the inability to detect total viable pathogen presence following disinfection. The research also investigates ways to optimize the use of current disinfection technologies to improve disinfection performance.

1.2 Chlorine Inactivation of Selected Pathogens in Treated Wastewater

C. parvum oocysts are extremely resistant to disinfection by FC and monochloramine, with CT values in excess of 7000 mg-minutes/L (Korich et al., 1990). Previous studies have shown that sequential chlorination with FC then monochloramine is more effective in inactivating *C. parvum* oocysts in water than either of them alone (Finch et al., 1995). When a dose of FC that produces no inactivation of *Cryptosporidium* oocysts is followed by a dose of monochloramine that produces only modest inactivation of *C. parvum* oocysts, a synergistic effect was observed with greater inactivation than either form of chlorine alone. A similar phenomenon was also reported for "in-situ" chloramination. Chlorination with FC followed by the addition of ammonium salts (e.g., after 1 minute) to form monochloramine ("in-situ" dynamic-chloramination) is much more effective for inactivating viruses than preformed monochloramine (Battigelli and Sobsey, 1994). Therefore, this study hypothesized that disinfection of treated sewage effluent by dosing FC that becomes converted to chloramines may produce some level of synergistic effects on certain types of microorganisms.

Previous studies have shown that enteric viruses and vegetative bacteria are inactivated relatively rapidly and extensively by FC but comparatively slowly and less extensively by monochloramine (Sobsey, 1989). Furthermore, aggregation and association with other particles protects bacteria and viruses from inactivation by chlorine species. In a disinfection study with purified viruses, a small "persistent fraction" of the viruses (about 1%) were found to be resistant to monochloramine even after re-dosing of the reaction mixture with either more viruses or more monochloramine (Venczel et al., 1992). Subsequent electron microscopic observation suggested that these more resistant viruses are indeed aggregated and therefore, more protected from monochloramine. Protection against chlorine inactivation by aggregation of bacteria was also demonstrated in previous studies. For example, in one study aggregated *Acinetobacter* was more than 30 to 40 times more resistant than dispersed *Acinetobacter* (Olson and Stewart, 1990). Similarly, the protective role of microbe association with solids, especially microbial occlusion in organic solids, was demonstrated to protect against inactivation by chlorine species. Studies have shown that cell-associated hepatitis A virus (HAV) was much more resistant to FC and somewhat more resistant to monochloramine than dispersed HAV (Sobsey et al., 1991) and that enteric bacteria within protozoans and crustaceans and nematodes are much more resistant to chlorine disinfection than freely suspended enteric bacteria (Chang et al., 1960; King et al., 1988; Tracy et al., 1966). Therefore, it appears that aggregation and association with organic particles may play a significant role in protecting microorganisms from wastewater chemical disinfection processes (Stewart and Olson, 1996). However, these phenomena have not been adequately explored for other new and emerging enteric microbes in wastewater.

1.3 UV Inactivation of Selected Pathogens in Treated Wastewater

UV irradiation has been studied for inactivation of some bacteria, viruses, and protozoans in wastewater and water (Sobsey, 1989), and it was shown to be effective against most of the microbes tested. UV doses of 5 mJ/cm^2 to 50 mJ/cm^2 reduce vegetative bacteria in water and treated wastewater by 90% to 99.9% (Chang et al., 1985; Qualls et al., 1985). For enteric viruses, 99.9% reductions require UV radiation doses of 20 mJ/cm^2 to 150 mJ/cm^2, depending on water quality and virus type (Battigelli et al., 1993). Although protozoan cysts and oocysts were considered very resistant to UV disinfection, recent studies (including research performed as part of this project) suggest that at very low doses, UV radiation is effective in inactivating many pathogenic protozoans like *C. parvum* oocysts (Clancy et al., 2000; Shin et al., 2001).

1.4 Mechanisms of Microbial Inactivation

1.4.1 Cryptosporidium

The mechanisms of chlorine inactivation of *Cryptosporidium* oocysts are unknown. However, due to the synergistic inactivation produced by sequential exposure of *Cryptosporidium* oocysts to FC and then monochloramine, it has been suggested that FC and monochloramine may react differently with *Cryptosporidium* oocysts to produce inactivation. Conversely, the mechanisms of UV inactivation of *Cryptosporidium* oocysts have been well developed over the past 3 years (e.g., Oguma et al., 2001; Zimmer et al., 2003). It is believed that nucleic acids are the primary targets for UV radiation, as is the case for bacteria and viruses.

1.4.2 Enteric Viruses

The mechanisms of chlorine inactivation for enteric viruses have been studied for some time, but there is still no consensus on the primary target: either the protein coat (capsid) or nucleic acid (ribonucleic acid [RNA] or deoxyribonucleic acid [DNA]). Several studies have reported that FC inactivates viruses primarily by altering the nucleic acid (O'Brien and Newman, 1979; Olivieri et al., 1975). However, Tenno et al., (1979) reported that hypochlorous acid (HOCl) inactivated poliovirus by reacting with the viral capsid and that the infectivity of the nucleic acid was not reduced. Young and Sharp (1985) also reported that the capsid alterations associated with virus inactivation were conformational changes. Furthermore, Alvarez and O'Brien (1982) suggested that the mechanism of action of FC is dose related, with inactivation at low doses due to subtle changes that do not greatly alter virion structure; however, at high doses major structural changes result in degradation of virions and release of nucleic acid. Virus inactivation by chloramines has also been suggested to be due primarily to alterations of viral capsid (Fujioka et al., 1983; Jacangelo and Olivieri, 1975), with sulfhydryl groups being the main targets for oxidation.

Mechanisms of UV inactivation of viruses have also been examined. As with bacteria, the primary target of UV in viruses is the nucleic acid. UV disinfection has demonstrated nucleic acid damage of both the RNA and DNA of viruses, both double- and single-stranded, including pyrimidine dimer formation, pyrimidine (6-4) pyrimidone photoproducts, and strand breaks (Furuse et al., 1967; Jagger, 1967; Winkler, 1964).

1.4.3 Enteropathogenic Bacteria

The mechanisms of chlorine inactivation of enteropathogenic bacteria are due to action of chlorine on the permeability of the outer cell envelope and oxidation of the sulfhydryl groups of proteins, including cell membrane proteins, enzymes, and other critical proteins (Stewart and Olson, 1996). However, some evidence suggests that nucleic acid damage may result, with possible formation of mutations (Camper and McFeters, 1979, Water Environment Federation, 1996). The mechanism of UV inactivation of enteropathogenic bacteria is due to direct damage of the cellular nucleic acids responsible for replication (Jagger, 1967).

1.5 Mechanisms of Pathogen Resistance

1.5.1 Morphological Considerations

Most bacteria contain plasmids, small double-stranded DNA molecules that exist and replicate separately from the chromosome. Plasmids contain genes whose protein products can confer important properties on an organism, such as resistance to antibiotics, heavy metals, and other inhibitory compounds, or the production of toxins or catabolism of specific substrates, such as pesticides and aromatic compounds. Plasmids in some strains of pathogenic bacteria may enable these organisms to resist disinfection by UV or chlorination because the plasmids encode for pili and other phenotypic characteristics that confer greater resistance to disinfection.

Viruses as a class of enteropathogens differ significantly in their morphology, including types and sizes of nucleic acid genomes as well as different capsid structures. The importance of these differences on the ability of a virus to survive a disinfectant needs to be investigated. The small size of viruses (0.02 microns to 0.3 microns) makes them potentially easily penetrable to UV radiation. However, the structure of a virus, consisting of either a nucleic acid surrounded by a protein coat (capsid), or in some viruses an envelope (a lipid bilayer membrane) surrounding the nucleocapsid, may act as a partial shield against UV penetration to the genetic material. For example, in the tobacco mosaic virus RNA is five times more sensitive to far UV radiation when it is separated from the capsid (Jagger, 1967). Hence, the protein coat contributes to viruses being more resistant to UV disinfection than bacterial cells. UV intensity at the virus surface must be strong enough to penetrate through the capsid, into the nucleic acid material. In addition, it would be expected that the effects of UV on single-stranded RNA versus double-stranded DNA viruses would differ. The primary sites of viral inactivation by free and combined chlorine may be on the capsid, and viral nucleic acid may remain intact and functional following disinfection. However, the existing scientific literature is inconclusive about which of these main targets is responsible for the inactivating event. The study of inactivation of viruses that exhibit numerous morphological differences would provide clues to elucidate of the role of fundamental composition and morphology in virus response to disinfectants.

In large cells, such as protozoans, penetration of far UV into cells may encounter a significant barrier. In a sufficiently large cell, virtually all of the UV absorption may be cytoplasmic (Jagger, 1967). Any resistance of protozoans such as *Cryptosporidium* to chlorine and UV was thought to be a consequence of the protective outer layer (wall) of the oocyst, which is highly resistant to chlorine and relatively impenetrable by and probably absorptive of UV radiation. However, we now know that protozoans such as *Cryptosporidium* and *Giardia* are very susceptible to UV radiation (Shin et al., 2001; Linden et al., 2002). The outer layers of these microbes do not offer significant protection against UV irradiation.

1.5.2 Effects of Combinations of Disinfectants on Survival of Pathogens

The effects of disinfectants in series or in combination have been initially investigated for *Cryptosporidium* in water as described above. The same types of enhanced disinfection effects may also apply to disinfection against resistant viruses and bacteria. In addition to the combination of free and combined chlorine, it would be beneficial to determine the effects of the addition of chlorine added as an odor-control agent followed by UV disinfection at the end of the treatment train. Similarly, the combined effects of UV followed by monochloramine for distribution of treated wastewater in reclamation situations may result in enhanced inactivation of pathogens beyond that capable of occurring with either disinfectant alone. The differences in the effects of UV radiation from MP UV sources as compared to LP UV sources, namely the differences in higher versus lower intensity UV and emission of a polychromatic versus monochromatic germicidal radiation, respectively, may also result in differences in inactivation of some persistent pathogens. The broad-band UV sources may emit wavelengths that damage some of the secondary metabolic processes and aid in cell inactivation. Research into all of these synergistic inactivation scenarios has not been investigated thoroughly.

1.5.3 Occlusion of Pathogens from Disinfectants—Water Quality Characteristics

A pathogen may be either nonparticle associated (free swimming) or particle associated. If the target pathogen is non-particle associated, then the disinfectant is unimpeded in its attack of the microorganism. Upon reaching a particle harboring a target pathogen, the disinfectant must first pass through associated particles or floc before reaching the target cell membrane. If the pathogen target is enmeshed in a floc, clumped with other organisms, or coated in a slime layer, it may be shielded from a lethal dose of chlorine or UV radiation (Parker and Darby, 1995). Disinfection is severely compromised when particulate matter is present in effluent. Neither UV nor chlorine is able to adequately penetrate the floc to get at organisms associated with particles (Emerick et al., 1999; Loge et al., 1997; Stewart and Olson, 1992). Sobsey et al. (1991) showed that compared to dispersed HAV, the virus associated with cells needs up to 10 fold the dose of FC to achieve a four-log inactivation. Particulate matter absorbs UV intensity to such a great extent that UV is unable to penetrate even a few microns into a floc (Loge et al., 1997). Loge et al. (1997) also illustrated that coliform are associated with particulate matter, and for the purpose of water-quality regulations, the presence of coliform organisms within particulates is of concern. However, little is known about the association of pathogenic bacteria, viruses, and protozoans to particulate matter in wastewater.

1.5.4 Inability to Detect Total Viable Pathogen Presence

1.5.4.1 Repair and Recovery Mechanisms

Numerous organisms that initially were assumed to be inactivated by a disinfectant have been shown to resuscitate following exposure to optimal recovery conditions. This phenomenon has been illustrated with bacteria subject to either UV irradiation or chlorination. Reactivation following UV irradiation has been attributed to photoreactivation or dark repair (Harm, 1980). Lindenauer and Darby (1994) have shown that the ability of coliform organisms to photoreactivate is directly related to the magnitude of the UV dose applied and dose of photoreactivating light. Recovery of injured bacteria following chlorination has been illustrated by culturing organisms on enhanced media. The ability to recover has been associated with applied chlorine dose. These bacteria reactivation and recovery phenomena have been studied extensively.

1.5.4.2 Viable but Nonculturable Response

The enteropathogenic bacteria normally found in wastewater are routinely exposed to a variety of both lethal and sublethal stressors, including chlorine and UV irradiation. The result is that the bacteria enter into either or both of an injured or VBNC state. Injury may be defined as those cells that are unable to grow in the presence of selective agents incorporated into the media commonly employed for their detection and enumeration. Such cells can be resuscitated by allowing their cellular repair by incubation on nonselective media at favorable temperature and pH. McFeters (1990) and Singh and McFeters (1990) have reviewed injury of indicator bacteria. LeChevallier et al. (1985) showed chlorine at 0.25 mg/L to 1.60 mg/L results in a 30% to 50% lethality and >90% "injury" to cells of *S. typhimurium*. In contrast, cells in the VBNC state (see below) are unable to grow in or on any medium, unless first allowed to resuscitate to the metabolically active state in a non-nutrient environment. This state is generally induced by the same stressors that are known to result in cell injury, but also by natural environmental stressors, such as low temperature or elevated osmotic concentrations. The presence of injured and/or VBNC cells leads to an underestimation of the true number of cells present, as such cells will not grow on the media routinely used for their detection. Thus, the extent of any microbiological health hazard present in wastewater cannot be accurately estimated.

To date, while a significant body of literature has developed on the production of injured cells by various chemical disinfectants (e.g. chlorine), little is known of the ability of UV radiation to cause cell injury. In either case, the great bulk of the work which has been performed on injured cells has been on cells in seawater or drinking water, and not on the enteropathogens present in wastewater. Desmonts et al. (1990), however, reported that up to 31.5% of the *Salmonella* spp. enumerated from chlorine-treated wastewater were in the VBNC state. Further, Camper and McFeters (1979) showed that after chlorination *Escherichia coli* was capable of resuming growth. Singh et al. (1986) found that such cells demonstrated virulence in the mammalian gut. Almost nothing is known about the role chlorine or UV irradiation play in inducing the VBNC state in enteropathogenic bacteria present in wastewater, or of the risk such cells may pose to the public.

Many bacteria, including *E. coli* and *Salmonella* spp. as well as many other human pathogens, may enter into the VBNC state. In this dormant state, the organisms can not be cultured by standard bacteriological methods, but retain viability, and can resuscitate back to the actively metabolizing and virulent state upon removal of the inducing stress. The VBNC state occurs in all gram-negative bacteria examined to date, and is induced by such environmental stresses as low and high temperature, elevated osmotic concentration, nutrient starvation, and even visible light. At least 17 genera of bacteria, comprising some 33 different species, are now known to enter the VBNC state (Oliver, 1993, 1995).

1.5.4.3 Resuscitation of Cells from the VBNC State

Resuscitation may be defined as a reversal of the processes that induced nonculturability, resulting in the ability of the cells to be culturable. It is clear, at least in the case of *V. vulnificus*, that resuscitation from the VBNC state occurs not only in the laboratory, but in both in situ and in vivo environments. As such an exit from the dormant state is critical to renewed metabolic activity and the subsequent reproduction of bacterial cells, studies on the conditions required for resuscitation are critical to any study on the role of the VBNC state in the ecology and epidemiology of *Salmonella* strains.

The effects of UV irradiation and chlorine treatment on induction of the VBNC state in bacteria present in wastewater systems needs to be examined. It is important to evaluate whether such disinfectants are effective in reducing the number of viable cells of enteropathogens or simply in inducing dormant states from which the cells may subsequently resuscitate. Ultimately, such studies should assay the potential health risk presented by wastewater that has been disinfected by such means.

1.6 Research Hypotheses and Objectives

The first premise of this research is that inactivation of pathogens in wastewater is not optimized because not enough is known about the mechanisms of inactivation and resistance of bacterial, viral, and protozoan pathogens. The mechanism of inactivation is impacted significantly by type of disinfectant (species of chlorine, type of UV source), the respective concentrations of disinfectant and contact times making up the total (chlorine or UV irradiation) disinfectant dose, and the physical structure of the pathogen of interest (microbe structure and composition and environmental adaptations). In addition, water quality (chemical species, UV transmission, particle presence) may severely inhibit the ability of a disinfectant to deliver a lethal dose to a pathogen or may act to completely protect pathogens from disinfectants.

The second premise of this research is that the phenomenon of pathogen recovery following disinfection is poorly understood. The reactivation potential of an organism exposed to a disinfectant is related to the degree of lethality of the dose received (which is a function of aggregation, particle association, and water quality), the repair mechanisms present (e.g., photoreactivating enzymes), and the environmental conditions following disinfection (temperature, nutrients, light, etc.). Furthermore, the ability to detect pathogens surviving disinfection is severely compromised by the inability to enumerate all viable microorganisms by traditional culturing techniques. Viable organisms may be temporarily injured subject to recovery on nonselective media or may be in the VBNC state.

The specific hypotheses guiding this investigation were:

1. Differences in morphology and composition affect the inactivation response of pathogens to disinfectants.
    a. Bacteria: bacteria with certain virulence and related properties, such as plasmids carrying virulence genes and antibiotic resistance factors and capsules, differ in their resistance to chlorine and UV technologies and in their ability to be reactivated or resuscitated,
    b. Viruses: viruses with different types and sizes of nucleic acid genomes as well as different capsid structures will differ in their ability to be inactivated with UV technologies or chlorine.
    c. Protozoans: protozoans such as *Cryptosporidium* are of a large enough size and complexity that they exhibit great resistance to single disinfectants, although disinfectants in combination or series may result in increased inactivation.
2. Altering ratios of disinfectant concentration (or intensity) and time, within identical total CT (or IT) values affects the inactivation effectiveness of chlorine and UV technologies.
3. Application of disinfectants in combination, in series, or repeatedly (step-feed disinfection) will enhance the ability to inactivate pathogens otherwise resistant to single disinfectant applications.
4. Sub- or semi-lethal disinfection may produce the appearance of inactivation although the cells are actually either just temporarily injured or in the VBNC state.
5. Pathogens of concern are aggregated or associated with particulate matter, and this aggregation or particle association interferes with adequate disinfection.

The overall goal of this research was to develop a better understanding of the mechanisms of action of, and pathogen resistance to, UV and chlorine inactivation for bacteria, viruses, and protozoans. The specific objectives of this research were to:

1. Elucidate the mechanism of action of UV and chlorine disinfectants on selected enteropathogens and model organisms
2. Define the characteristics that certain pathogenic organisms have that make them relatively more resistant than others to UV and chlorine disinfectants
3. Determine the effect of variations in concentration and time factors at equal disinfectant doses on pathogen inactivation and subsequent recovery
4. Evaluate the effect of combinatorial or sequential application of disinfectants on the inactivation of persistent pathogens
5. Investigate induction of VBNC responses resulting from exposure to UV and chlorine disinfectants
6. Determine the extent to which pathogenic organisms are aggregated and associated with particulate matter utilizing genetic probes and physical separation methods, and the subsequent impact this association has on pathogen inactivation.

1.7 Research Approach

The research focused on examining two types of disinfectants: UV radiation (from LP and MP UV sources) and chlorine dosed as FC or monochloramine or dosed as FC that gets converted to chloramines by reaction with ammonia in wastewater. Within this framework, the sequential use of two disinfectants was also examined. This sequential disinfection was employed to compensate for the lack of efficacy of one type of disinfectant against a class of microbes by using another type of disinfectant that is highly efficacious against that class of microbe. The possibility of synergism brought about by the use of the two different disinfectants was also investigated because it is a phenomenon reportedly observed in previous disinfection studies of water- and wastewater-borne microbes.

Different types of test waters were studied, including reagent grade lab waters and wastewaters that differed in water quality and particle concentration and size. These waters were investigated because water quality can influence microbial inactivation. For example, water quality, such as pH or ammonia concentration, will affect the speciation of chlorine in wastewater, affecting the capability for inactivation. Particles in wastewater were investigated for their effect on microbial inactivation. Particle-association, especially microbe embedded particles and/or aggregated microbes, are protected from various disinfectants, according to some previous studies. Investigating the role of particle-association on microbe inactivation required the use of reliable methods to ascertain and quantify particle association of microbes. Because these methods are not well developed and standardized, they had be investigated and standardized as part of the research effort.

Various types of wastewater-borne microbes were investigated to provide a range of application of the research. Viruses, bacteria, bacterial spores, and protozoans were investigated, all of which are of concern when they are present in drinking water, wastewater effluent, and recreational water. The ability of some of these test microbes to repair UV-induced DNA damage was also studied. This is a fundamental property of many cellular microbes and must be considered when using UV for disinfection. In addition to repair UV damage, the ability of injured but still viable and therefore potentially infectious bacteria to go undetected after exposure to disinfectants, due to the limitations of the culture methods employed, was investigated. Such injured bacteria may be detected by resuscitation culture methods, thereby giving a better estimate of the true extent of bacterial inactivation that results in death, as opposed to non-culturability.

This report is divided into sections describing different aspects of the work, including:

- Inactivation and repair of pathogens with UV irradiation
- Investigating *Cryptosporidium*
- Viruses and the effectiveness of different wavelengths for inactivation
- The effect of UV and chlorine disinfectants on induction of enterobacteria into the VBNC state and potential for resuscitation
- Mechanisms of inactivation and resistance
- Investigating the effect on UV and chlorine disinfection of aggregation of viruses and bacteria
- The association of microbes with particles and the detection of microbes associated with particles
- Optimizing application of disinfectant doses for pathogens in wastewater with varying concentration and time factors and evaluating sequential disinfection using combinations of UV and chlorine

CHAPTER 2.0

INACTIVATION OF PATHOGENS WITH UV IRRADIATION

2.1 Introduction

C. parvum is an important health-related waterborne pathogen that is ubiquitous in surface and source waters (LeChevallier and Norton, 1995) and very resistant to conventional water treatment processes (Gibson et al., 1998). Although conventional water filtration systems achieve some removal (Gibson et al., 1998), chemical disinfection by chlorination is incapable of achieving appreciable inactivation of *C. parvum* oocysts at practical disinfectant doses and contact times (Clark et al. 1993; Sobsey, 1989). Previous studies using the in vitro viability assays of excystation and vital dye staining suggested that *C. parvum* oocysts are also very resistant to monochromatic LP UV radiation (Campbell et al., 1995; Ransome et al. 1993), and much more resistant than the enteric viruses that have been proposed as the basis for determining UV dosimetry in water and wastewater treatment (Havelaar, et al. 1990; Havelaar, et al. 1991; Nieuwstad et al. 1991). The inactivation of *C. parvum* oocysts by UV irradiation from both LP and MP UV lamps was determined along with an assessment of the wavelength dependence of UV inactivation. As a benchmark, the inactivation of *C. parvum* oocysts was compared to coliphage MS2, a widely used viral indicator of UV radiation disinfection efficacy in water.

In addition to *C. parvum*, other waterborne pathogens are considerably resistant to UV irradiation, especially some important human enteric viruses (Chang et al., 1985; Meng and Gerba, 1996). These enteric viruses are reported to be 10 to 100 times more resistant to UV irradiation than enteric bacteria and parasites. Therefore, viruses will likely be the limiting organisms in determining UV reactor design and UV doses for systems requiring waterborne pathogen inactivation.

There are more than 100 enteric viruses that could be transmitted by water (Melnick and Gerba, 1980). They are different in terms of size, genomic material, protein capsid, and presence or absence of an outer envelope. However, reliable data on virus inactivation by UV irradiation are limited to only several enteric viruses and a few bacteriophages, and there seems to be a wide spectrum of sensitivity to UV irradiation by those viruses (Sobsey, 1989, Chang et al., 1985, Meng and Gerba, 1996). With more enteric viruses and other microorganisms emerging as newly recognized waterborne pathogens, it would be beneficial to determine the major factors influencing virus sensitivity to UV irradiation. Therefore, several waterborne pathogenic viruses and bacteriophages were selected with different sizes and genomic material: three small RNA viruses (poliovirus 1 [PV1], Coxsackievirus B4, and bacteriophage MS2) and three relatively large DNA viruses (Adenovirus 2, and two DNA bacteriophages, members of the Podoviridae and Myoviridae families, respectively). The inactivation by monochromatic, LP UV was determined. The purpose of studying these representative viruses was to determine the relationships of virus size and genomic material to their sensitivity to UV irradiation and to determine if there is a predictable pattern of UV sensitivity relative to virion or genomic properties.

Water and wastewater quality factors, such as pH, temperature, inorganic (and organic) matter, and particulates, greatly influence disinfection efficiency (Sobsey, 1989). Unlike other chemical disinfectants, however, UV disinfection is relatively pH and temperature independent. Nonetheless, efficacy of UV radiation is greatly influenced by dissolved organic matter and particles in water and wastewater that absorb and/or scatter UV irradiation to affect the delivery of UV irradiation to target microorganisms. Therefore, in order to better define the practical UV doses for effective inactivation of *C. parvum* oocysts in wastewater, the inactivation of *C. parvum* oocysts by UV radiation in not only laboratory reagent-grade water but also wastewater was determined.

2.2 UV Inactivation of *C. parvum* oocysts and Coliphage MS2

2.1.2 Materials and Methods

2.1.2.1 Microbial Preparation

C. parvum oocysts (Iowa strain) were purchased from Pat Mason, Pleasant Hills Farm (Troy, Idaho). Shed oocysts collected daily from experimentally infected 3-day old calves were screened to remove large debris and hair and then purified and dispersed by processing through discontinuous sucrose gradients, followed by cesium chloride gradients (1.15 g/ml, 1.15 specific gravity). Recovered oocysts were washed in phosphate buffered saline (PBS, pH 7.2), resuspended in PBS containing 1000 units penicillin/mL and 1000 mg streptomycin/mL and stored at 4°C until use (usually within 2 months). Bacteriophage MS2, a male-specific (F+) RNA coliphage, was grown and assayed in *E. coli* C3000 by the double agar layer (DAL) plaque technique (Havelaar et al., 1990; Havelaar et al., 1991; Nieuwstad and Havelaar, 1994). Virus stock was produced by scraping the top agar layer of DAL plates having confluent lysis of host cells into a small amount of PBS, extracting with an equal volume of chloroform, centrifuging at 4000 X g for 30 minutes at 4°C, and recovering the supernatant fluid. These preparation methods have been shown to be effective to make relatively monodispersed microbial suspension in previous studies based on observation with hemocytometer (*C. parvum* oocysts) or by transmission electron microscopy (Coliphage MS2).

2.1.2.2 LP and MP UV Systems

The LP collimated beam UV apparatus consisted of two 15-watt lamps that produced UV radiation at 254 nm. The radiation was collimated through a circular opening to provide perpendicular radiation to the surface of the test suspension (0.255 cm thickness for 5 mL sample) in a 60 X 15 mm cell-culture petri dish. The MP UV system used for this study was a collimated beam, bench-top irradiation system with one MP UV lamp (Rayox, 1KW lamp, provided by Calgon Carbon Corporation [Pittsburgh, Pennsylvania]), housed in a reflector. The radiation was "collimated" through a 2" diameter circular tube to the surface of the test suspension in a 60 X 15 mm cell-culture petri dish. UV irradiance from the MP UV system was monitored with a radiometer (IL1770, SED240 detector, G filter, W diffuser) that provided direct measure of germicidal UV energy weighted to the German Deutsche Industrialen Normalen (DIN) and Illumination Engineering Society (IES) germicidal effectiveness curves, housed in a reflector. The generic set-up of a collimated beam bench scale UV apparatus is illustrated in Figure 2-1.

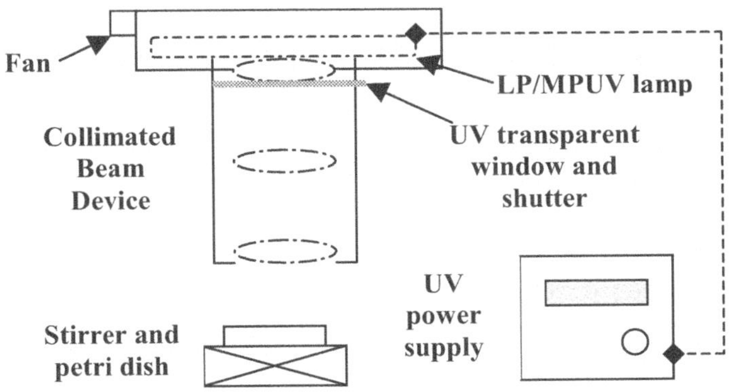

Figure 2-1. Laboratory Apparatus for UV Collimated Beam Disinfection.

2.1.2.3 UV Dose Measurement

UV irradiance from LP UV was measured at the University of North Carolina–Chapel Hill (UNC–CH) with an International Light IL500 radiometer at a wavelength of 254 nm. The radiometer used for LP and MP UV experiments at Duke University (International Light, IL1770, SED240 detector, W diffuser, G filter) was recalibrated to responsivity determined at 2 nm increments from 200 nm to 310 nm. The filter used with the radiometer is a specially designed germicidal filter that internally weighs the detection of a polychromatic spectra to provide direct measurement of effective germicidal radiation, spectrally weighted to match the IES Luckiesh and German DIN standard germicidal effective irradiance curves. This radiometer can, in theory, be used to read the effective germicidal radiation from a polychromatic source such as the MP UV. When measuring UV irradiance with a monochromatic UV source, the sensitivity factor for UV-254 nm is enabled to provide calibrated irradiance readings. The IL1770 radiometer was used with the germicidal filter enabled, thus all radiometer irradiance measurements and subsequent radiometer doses represent the weighted germicidal effectiveness spectra. An estimation of the actual UV irradiation in the 220 nm to 300 nm range can be obtained from the detailed calibration data along with the emission spectra of the MP UV lamp using a spreadsheet (Linden and Darby, 1997).

The measured incident irradiance values at the surface of the test liquid were corrected for any nonhomogeneity of irradiation across the surface area of the petri dish to provide a value for the average incident irradiance. This "petri factor" was determined for each mode of operation for the collimated beam (CB) units. Spreadsheets for calculation of the petri factor and determination of average incident irradiance across the surface area of the petri dish were developed to facilitate the determination of exposure time necessary to achieve a given UV dose. The average irradiance in the mixed suspension was determined mathematically from an integration of the Beer Lambert Law over the sample depth, accounting for UV absorbance and incident average irradiance (Morowitz, 1950). Target doses were computed as the product of average irradiance and time (in seconds), and required exposure times were calculated by dividing the desired UV dose by the average UV irradiance.

2.1.2.4 Experimental Protocol for UV Disinfection Experiments

Studies were performed in both lab water (PBS) and wastewater. Five-milliliter volumes of sample containing purified *C. parvum* oocysts or coliphage MS2 at concentrations of ~10^6 organisms/mL in 60 X 15 mm cell culture petri dishes were irradiated with a UV CB system while magnetically stirring the samples slowly (~30 rpm) at room temperature (23°C to 25°C). After predetermined exposure times corresponding to target UV doses, samples were removed from the UV irradiation system and diluted serially 10-fold for subsequent infectivity assays.

Wastewater samples were taken from Southern Durham Water Reclamation Facility (WRF; Durham, North Carolina). This facility has primary physical, secondary biological, and tertiary filtration processes. The tertiary filtrate is disinfected with LP UV. Samples were taken before and after the UV disinfection unit, and a known concentration of *C. parvum* oocysts was spiked into the wastewater sample for bench-scale UV disinfection experiments.

2.1.2.5 *C. parvum* Infectivity Assay

C. parvum infectivity assays were done in Madin-Darby Canine Kidney (MDCK) cell cultures (ATCC CCL 34). MDCK cells were maintained in Dulbecco's Modified Eagles Medium (DMEM/F1) supplemented with 10% fetal bovine serum, 2 mM L-glutamine, and 15 mM HEPES. For assays, cells were seeded at 7.5 x 10^4 cells per double-chambered Lab Tech Chamber Slide (Nalge Nunc International, Rochester, New York) in a volume of 1.5 mL of Ultraculture™ serum-free medium (BioWhittaker, Inc., Walkersville, Maryland) supplemented with 250 µg/mL kanamycin, 50µg/mL gentamicin, 150 µg/mL mycostatin, and 2 mM L-glutamine. Cultures were incubated at 37°C in a 5% CO_2 environment for 4 days. Four-day-old MDCK cell monolayers in slide chambers were inoculated with duplicate 300-µl volumes of undiluted or serially 10-fold diluted oocyst suspensions, and incubated for 3 hours at 37°C to promote excystation and initial cell infectivity. After incubation, the cell monolayers were washed with PBS, 1.5 mL of fresh Ultraculture medium was added to the chambers, and the cells were incubated for 48 hours. After incubation, cell layers were washed with PBS, fixed in Bouin's solution (Sigma, St. Louis, Missouri), decolorized, and blocked with 1% bovine serum albumin (BSA) in PBS. Cell layers were inoculated with 300 µl of C3-C3 antibody specific for *C. parvum* living stages that had been conjugated with the red fluorochrome Cy3 or with apple green fluorescein isothiocyanate. Antibody-treated cell layers were incubated for 90 minutes in the dark at room temperature, after which the labeled cell cultures were washed, mounted under coverslips and observed microscopically. Labeled life stages were assessed by UV-illuminated epifluorescent microscopy using either a 25X or 40X objective. Infectivity was assayed by a quantal approach in which the presence or absence of living stages (sexual gamonts and asexual meronts) was scored in 50 to 100 sequential, non-overlapping fields. Any field containing a fluorescent developmental stage of *Cryptosporidium* received a positive score. A negative score was assigned to those fields that contained no fluorescent life stages. The number of infected and uninfected test units fields was scored and recorded. Oocyst infectivity titer was calculated as a most probable number (MPN) from the Thomas equation using the dilutions where only some of the fields observed were infected.

2.1.2.6 Data Presentation

The titers of bacteriophage MS2 were calculated as plaque forming units (PFU) per milliliter based on plaque counts in duplicate plates. The titers of *C. parvum* were calculated as MPN per milliliter based on counts in ~50 to 100 sequential, nonoverlapping fields. For each microorganism, the concentrations in the control sample were computed and taken as N_o, the initial microbe concentration. For each test sample, the average concentration of each microorganism was computed. The proportion of initial microorganisms remaining at each dose (D) was computed by dividing the concentration at each dose (N_D) by the initial concentration (N_o). These values were then $\log_{10}$-transformed ($\log_{10}(N_D/N_o)$), and the values of replicate experiments were averaged. The data for $\log_{10}(N_D/N_o)$ of either individual experiments or averaged values of replicate experiments were then paired with the data for dose (D) and plotted. The number of replicate experiments varied for each UV dose, but more than two replicate experiments were done for most of the doses. The ranges of inactivation ($\log_{10}$ inactivation) values is given in the legend in the corresponding figures for a series of experiments (see Section 2.1.3).

2.1.3 Results

2.1.3.1 LP UV Radiation Results

Figure 2-2 shows the average reduction kinetics of *C. parvum* and MS2 by several different doses of LP UV radiation in PBS at room temperature. The reductions of *C. parvum* infectivity were very rapid, approximately first-order, and reached the detection limits of the infectivity assays (~3 $\log_{10}$) within a dose of 3 mJ/cm^2, similar to that found by Clancy et al. (2000). The reductions of coliphage MS2 were considerably slower and less extensive than those of *C. parvum*, with only a ~2 $\log_{10}$ reduction at a dose of 30 mJ/cm^2. The MS2 reduction kinetics by LP UV radiation were similar to those previously reported in the literature (Meng and Gerba, 1996; Nieuwstad and Havelaar. 1994; Tree et al., 1997; Wilson et al., 1992).

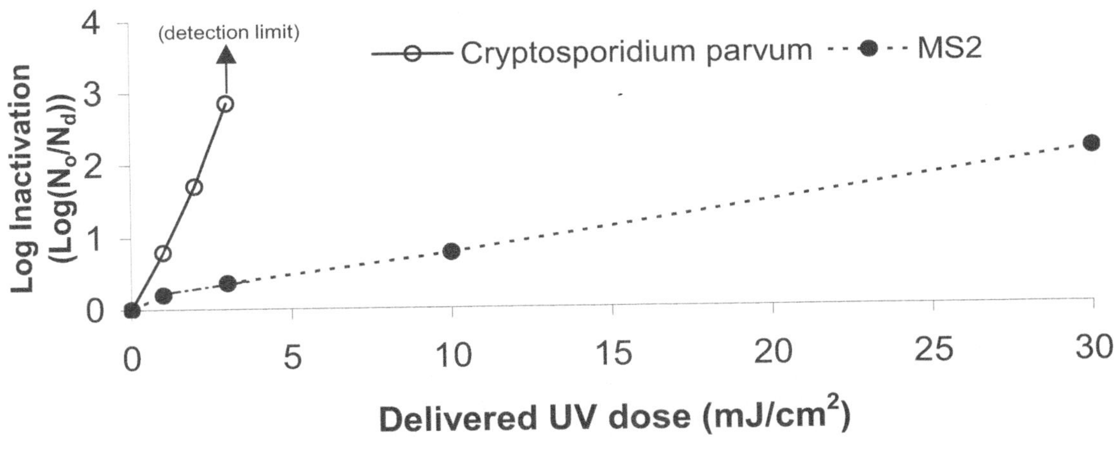

Figure 2-2. Inactivation Kinetics of *C. parvum* and coliphage MS2 by Monochromatic, LP UV Radiation. Data ranges for replicate experiments ($\log_{10}$): *C. parvum* oocysts (1 mJ/cm^2: 0.6 to 1.0; 2 mJ/cm^2: 1.7 to 1.7; 3 mJ/cm^2: >2.85), coliphage MS2 (1 mJ/cm2: 0.1 to 0.3; 3 mJ/cm^2: 0.3 to 0.5; 10 mJ/cm^2: 0.7 to 0.8, 30 mJ/cm^2: 2.1 to 2.1).

2.1.3.2 MP UV Radiation Results

The average reduction kinetics of inactivation of *C. parvum* oocysts and MS2 by several different doses of polychromatic, MP UV radiation in PBS at room temperature are illustrated in Figure 2-3. As with monochromatic LP UV radiation, the reductions of *C. parvum* infectivity were very rapid and reached the detection limits of the infectivity assays (~3 $\log_{10}$) within a dose of approximately 3 mJ/cm^2. The observed rate and extent of inactivation of *C. parvum* oocyst infectivity appears to be more rapid than that reported by Bukhari et al. (1999), although it is difficult to compare the results here to those previous findings because the doses of the previous study are higher by as much as an order of magnitude (3 mJ/cm^2 and 40 mJ/cm^2).

The average reduction kinetics from replicate experiments of coliphage MS2 by polychromatic, MP UV radiation were considerably slower and less extensive than those of *C. parvum* oocysts. There was a gradual reduction of MS2 within the tested dose range and only a ~2 $\log_{10}$ reduction of MS2 was achieved at a dose of 30 mJ/cm^2 of MP UV radiation. These findings are consistent with those obtained fro LP UV disinfection. Higher doses of UV were not presented for MS2 because they were similar in previous studies (Battigelli et al., 1993) and because *C. parvum* oocyst infectivity has already been reduced to detection limits at the very low initial UV doses tested here.

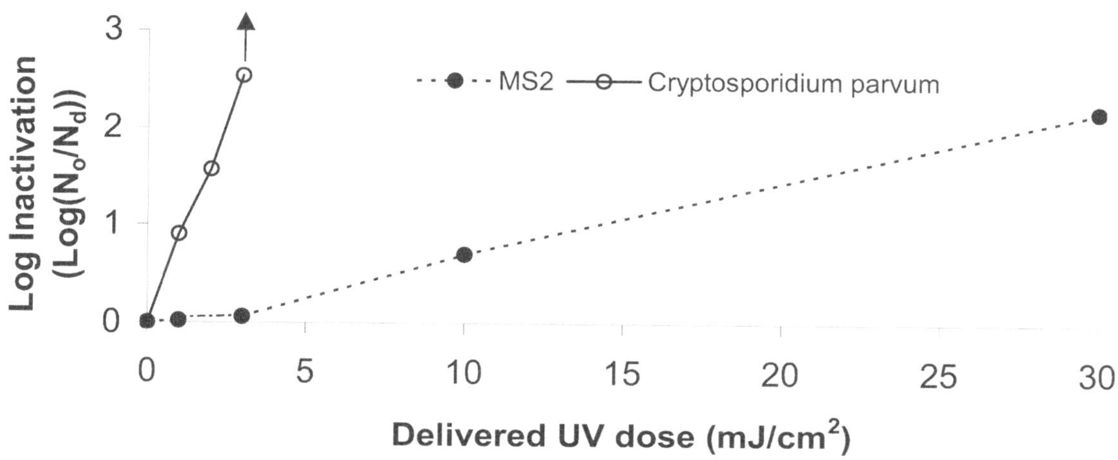

Figure 2-3. Inactivation of *C. parvum* oocysts and Coliphage MS2 by MP UV Radiation. Data ranges ($\log_{10}$): *C. parvum* oocysts (1 mJ/cm^2: 0.8 to 1.0; 2 mJ/cm^2: 1.6; 3.7 mJ/cm^2: >2.53), coliphage MS2 (1 mJ/cm^2: 0.02; 3 mJ/cm^2: 0.06; 10 mJ/cm^2: 0.7 to 0.8; 30 mJ/cm^2: 2.1 to 2.3).

2.1.3.3 Comparison of LP and MP UV Radiation Results

The average reduction kinetics of *C. parvum* infectivity by several different doses of monochromatic LP and polychromatic MP UV radiation applied in PBS at room temperature are illustrated in Figure 2-4. The reductions of *C. parvum* infectivity by both types of UV irradiation were very rapid and reached the detection limits of the infectivity assays (~3 $\log_{10}$) within a dose of 3 mJ/cm^2. The reduction kinetics by LP and MP UV irradiation were very similar. Although the reduction of MS2 was more gradual (only a ~2 $\log_{10}$ reduction at a dose of 30 mJ/cm^2), the reduction kinetics of coliphage MS2 by LP and MP UV irradiation were also very similar (Figure 2-5).

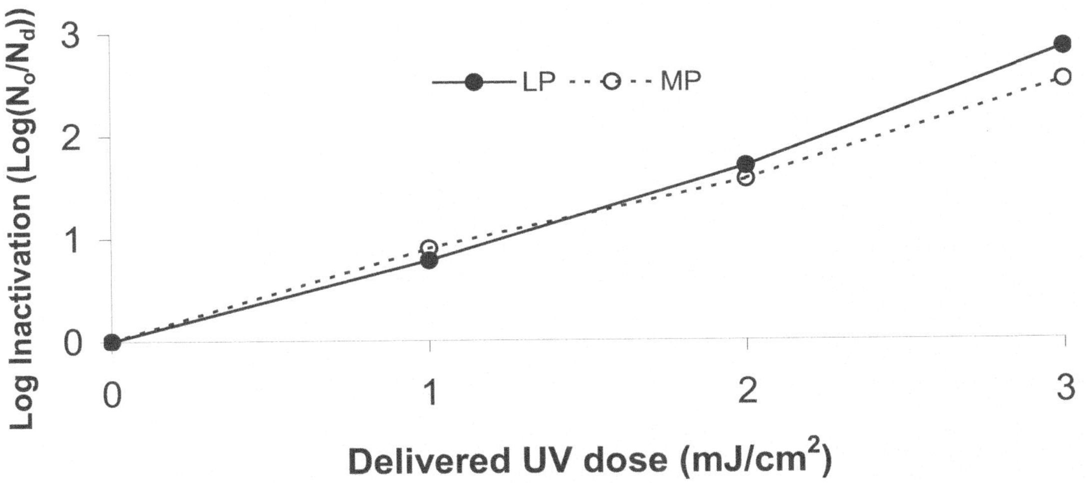

Figure 2-4. Inactivation Kinetics of *C. parvum* oocysts by LP and MP UV Radiation. Dose range based on Figures 2-2 and 2-3.

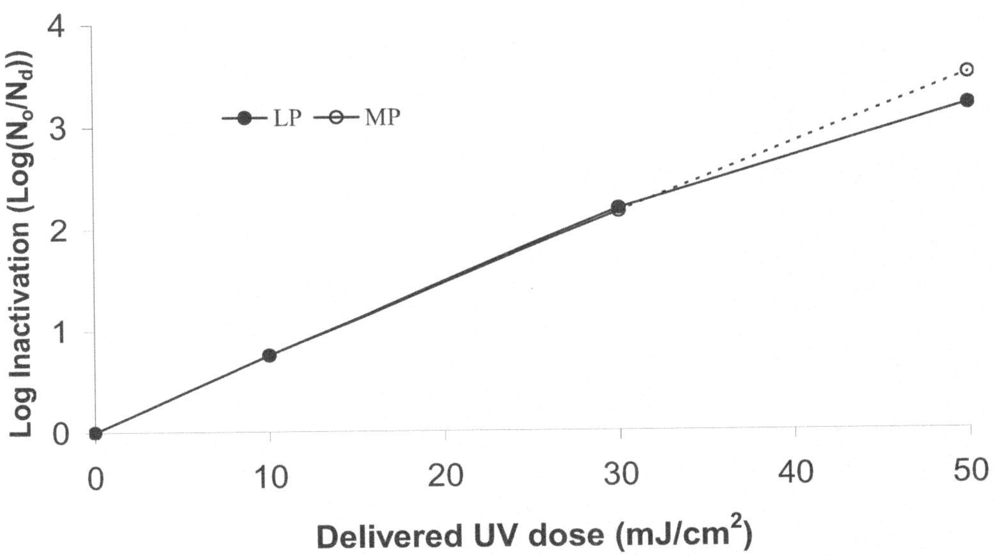

Figure 2-5. Inactivation Kinetics of Coliphage MS2 by Low- and Medium-pressure UV Radiation. Dose range based on Figures 2-2 and 2-3.

2.1.3.4 LP UV Radiation Results in Wastewater

Figure 2-6 shows the reduction kinetics of *C. parvum* oocyst infectivity by several different doses of LP UV radiation in wastewater at room temperature. The reduction of *C. parvum* infectivity was very rapid and reached the detection limits of the infectivity assays (>3.3 $\log_{10}$) at a dose of 3 mJ/cm^2. This inactivation kinetics is similar to those generated in PBS. This experiment was done only one time with one wastewater sample due to the time and personnel constraints. However, two lines of evidence support the validity of the findings. The results are very similar to those obtained in clean water, specifically PBS, as noted above. Furthermore, the results are consistent with the data on inactivation kinetics of *C. parvum* oocysts in another wastewater (see Sections 5.2 and 5.4). Overall, it appears that an appreciable extent of inactivation of *C. parvum* oocysts is achieved in wastewater within a practical range of UV dosage.

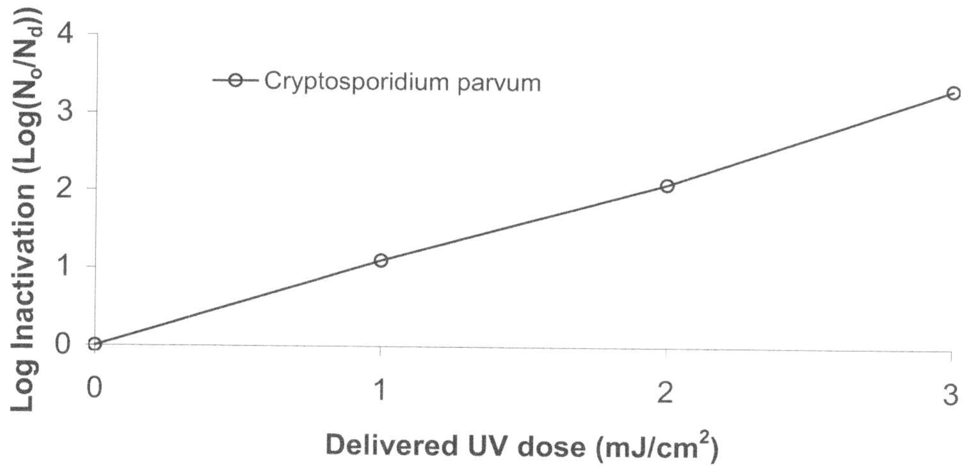

Figure 2-6. Inactivation of *C. parvum* oocysts in a Wastewater by LP UV Radiation.

2.3 UV Inactivation of Viruses

2.3.1 Materials and Methods

2.3.1.1 Virus Preparation and Assays

Poliovirus 1 (PV1, strain LSc) and Coxsackievirus B4 (CVB4, a field isolate from fecally contaminated water provided by Pierre Payment, Institute Armand Frappier, University of Quebec, Lavalle, Quebec, Canada) were propagated and assayed in confluent layers of buffalo green monkey kidney cells. Adenovirus type 2 (AV2) was propagated and assayed in confluent layers of A549 cells. Cells were maintained in Eagle's Minimum Essential Medium containing 2% heat-inactivated fetal bovine serum as described by Sobsey et al. (1978). Infected cell lysates were frozen and thawed three times and extracted by homogenizing in an equal volume of chloroform. The supernatant was recovered by low speed (5000 x g) centrifugation for 15 minutes at 4°C. F-specific coliphage MS2 was propagated and assayed in *E. coli* C3000 by the double agar layer plaque technique. Coliphage V1 and a Myoviridae that was a field isolate from ground water in Orange County, California, were propagated and assayed in *E. coli* strains WG5 and C, respectively, by essentially the same techniques as used for coliphage MS2. For virus stock, the top agar layer having confluent lysis of the host cells in double agar layer plates was harvested by scraping into a small amount of PBS, and the virus was extracted with an equal volume of chloroform. The supernatant was recovered by low speed (4000 x g) centrifugation for 30 minutes at 4°C. All virus extracts were further purified by centrifugal ultrafiltration (Centricon 100) and dispersed by filtration through microporous polycarbonate filters (0.2 µm for DNA viruses and 0.2 µm and 0.08 µm in series for RNA viruses) pre-treated with 0.1% Tween 80 solution in water. Previous studies have shown that these filtration procedures provide nearly monodispersed virus preparations as evidenced by >99.99% dispersed virions and very small aggregates (<9 virions per aggregate), as evidence by transmission electron microscopy using the kenetic attachment method to preserve the state of virus aggregation is suspension when preparing electron microscope specimens (Battigelli et al., 1993).

The UV dosimetry and experimental procedures were the same as reported for *Cryptosporidium*.

2.3.2 Results

Figure 2-7 shows the inactivation kinetics of the three RNA viruses by several different doses of monochromatic LP UV radiation in PBS at room temperature. The inactivation rates of the two animal RNA viruses, PV1 and CVB4, by LP UV were rapid, approximately first-order, and reached ~4 $\log_{10}$ and >5 $\log_{10}$, respectively, within a UV dose of 30 mJ/cm^2. However, the inactivation rate of the RNA bacteriophage, MS2, was much slower and only 2 $\log_{10}$ inactivation was achieved at the same UV dose of 30 mJ/cm^2.

Figure 2-8 shows the inactivation kinetics of the three DNA viruses by several different doses of monochromatic LP UV radiation in PBS at room temperature. The inactivation rates of the two double-stranded, linear DNA bacteriophages were very rapid and reached ~4 $\log_{10}$ and ~6 $\log_{10}$ with a UV dose of 10 mJ/cm^2. However, the inactivation of the similar sized linear, double-stranded DNA virus, AV2, was much slower. Only modest inactivation was observed within the tested dose range. For AV2, the reductions were ~2 $\log_{10}$ at a dose of 60 mJ/cm^2 and ~4 $\log_{10}$ at a dose of 120 mJ/cm^2.

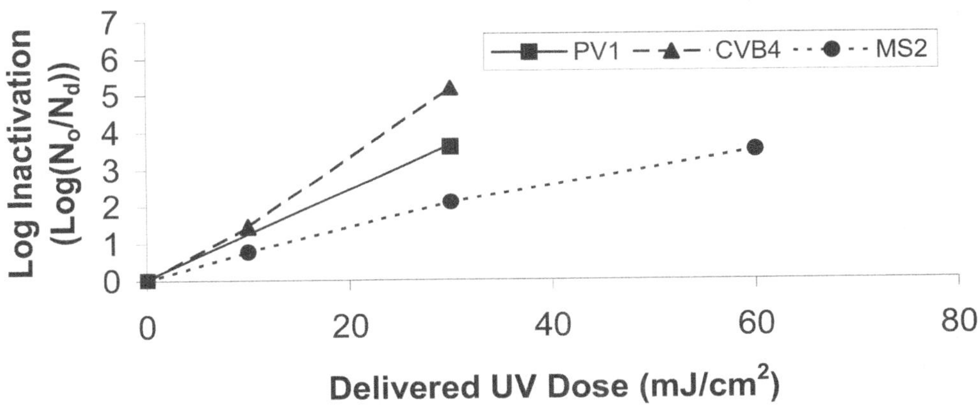

Figure 2-7. Inactivation of RNA Viruses by LP UV. Data ranges ($\log_{10}$): PV1 (30 mJ/cm^2: >3.6), CVB4 (10 mJ/cm^2: 1.3-1.6; 30 mJ/cm^2: >5.18), coliphage MS2 (10 mJ/cm^2: 0.7-0.8; 30 mJ/cm^2: 2.1; 60 mJ/cm^2: 3.5).

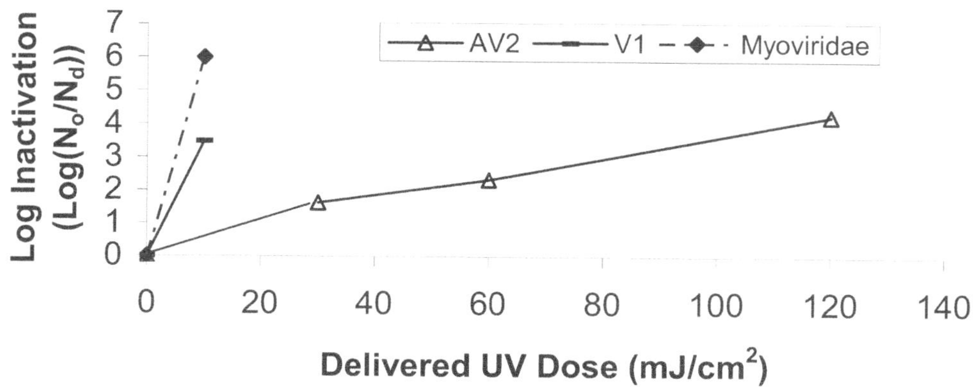

Figure 2-8. Inactivation of DNA Viruses by LP UV. Data Ranges (log_{10}): AV2 (30 mJ/cm²: 1.5-1.8; 60 mJ/cm²: 2.2-2.4; 120 mJ/cm²: >4.25), bacteriophage V1 (10 mJ/cm²: >3.48), Myoviridae (10 mJ/cm²: >5.99).

Results indicate that inactivation of human enteric viruses and bacteriophages by UV irradiation are not reliably predicted by the type and size of the virus or its nucleic acid genome. The two human pathogenic viruses, PV1 and CVB4 that are small (~27 nm) and have single-stranded, linear RNA genomes of about 7.5 kb, were relatively sensitive to LP UV radiation. However, the single-stranded, linear RNA bacteriophage with similar virion size and somewhat smaller genome size (about 4 kb), coliphage MS2, was relatively resistant to UV irradiation. In contrast, the two large (50 nm to 100 nm) DNA bacteriophages with double-stranded, linear DNA genomes were relatively sensitive to LP UV radiation. However, a human pathogenic DNA virus with similar size, and linear, double-stranded genome, AV2, was highly resistant to LP UV radiation. This suggests that there is no strong association between virion size and genetic composition of enteric viruses and their response to LP UV irradiation. It is possible that sensitivity of a virus to LP UV is determined by other chemical differences in nucleic acid genome composition (such as relative G + C content) as well as the composition and organization of amino acids, carbohydrates, and where relevant, lipid composition of the protein capsid. More detailed biochemical and molecular biological research is recommended to elucidate the roles of physical and chemical properties of viruses and their sensitivity to UV radiation.

2.4 Relative Efficacy of UV Wavelengths for *C. parvum* Oocysts and MS2 Coliphage

2.4.1 Introduction and Purpose

One question that arises when comparing the effectiveness of monochromatic and polychromatic UV sources for microbial inactivation is what is the advantage, if any, of using lamps that produce polychromatic UV light compared to those that emit monochromatic UV light? Early investigation (Gates, 1929) describing the relative efficacy of different wavelengths in the UV-C range (200 nm to 300 nm) for the inactivation of bacteria showed that wavelengths other than 254 nm are not only microbiocidal, some are more microbiocidal than the UV 254 emitted from LP lamps. However, little is known about the relative efficacy of UV wavelengths for inactivation of *C. parvum* oocysts and coliphage MS2. This information has important implications for understanding the underlying mechanisms of microbial inactivation by UV disinfection and for the subsequent design of such disinfection systems. Therefore, the efficacy of distinct bands of UV light for inactivation of *C. parvum* oocysts and coliphage MS-2 was determined.

2.4.2 Materials and Methods

The microbial methods, experimental protocol and infectivity assay for *C. parvum* and MS2 coliphage were the same as reported above.

2.4.2.1 Filtering Polychromatic UV Radiation

The same MP UV system described previously was employed using a series of UV bandpass filters. These UV bandpass filters (Andover Corp, Salem, NH) were used to elucidate the principal effective wavelength range for inactivation of *C. parvum* and coliphage MS2. A set of eight UV interference filters were mounted at the outlet of the collimating tube of the MP UV system for use one at a time such that all radiation incident on the surface of the liquid sample passed through the filter, as illustrated in Figure 2-9. Each filter had an approximate ½ peak bandwidth of 10 nm, centered on the following wavelengths: 214 nm, 230 nm, 240 nm, 256 nm, 261 nm, 270 nm, 280 nm, and 290 nm. *C. parvum* was exposed to doses of 2 mJ/cm^2 and 5 mJ/cm^2 while MS-2 was exposed to doses of 10 mJ/cm^2 and 30 mJ/cm^2 for each filter used. UV doses were corrected for the absorbance of the suspension at each wavelength, ensuring that a corrected average dose in a mixed suspension was delivered. The spectrum of radiation transmitted through each filter was verified using spectrometry (Spectral Instruments 440, Spectral Instruments [Tucson, Arizona]).

The UV dose of the filtered radiation energy emitted from the MP UV source was determined utilizing potassium ferrioxalate actinometry as described by Linden and Darby (1998). The quantum yield (moles of Fe^{+3} reduced to Fe^{+2} divided by the moles of photons absorbed) of potassium ferrioxalate is well known throughout the UV range. The absorbance of radiation by the actinometer in the UV-C and UV-B range is above 99%, making it well suited to measure distinct bands of UV radiation. In addition to the actinometric measures of UV dose, the irradiance of radiation through each filter was monitored utilizing the radiometer described above. The radiometer was calibrated with responsivity determined at 2-nm increments from 200 nm to 310 nm. Thus, as each bandpass filter was monitored, the radiometer sensitivity factor was adjusted to display absolute irradiance values for the peak transmission line of that specific bandpass filter. This radiometer reading was used to control irradiance and dose during exposures with test microbes. Characteristics of the bandpass filters are presented in Table 2-1.

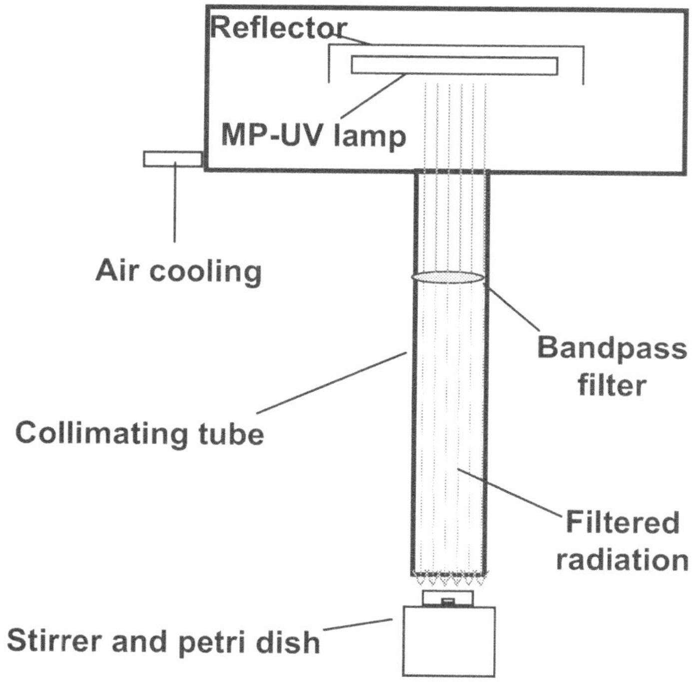

Figure 2-9. MP UV Collimated Beam Apparatus with Bandpass Filter in Place.

Table 2-1. Characteristics (nm) of Bandpass Filters.

Filter	Peak wavelength	1/2 peak bandwidth
214	216	209-219
230	230	225-235
240	242	235-245
256	255	251-261
261	263	256-266
270	271	265-275
280	281	275-285
290	290	285-295

2.4.3 Results of Bandpass Filter Study

Figure 2-10 shows the reduction of *C. parvum* oocyst infectivity resulting from exposure to UV at a dose of 2 mJ/cm^2 as a function of the peak throughput for each filter. Similarly, Figure 2-11 shows the results for the MS-2 exposures for a dose of 30 mJ/cm^2. These two UV doses were chosen because they represent UV doses in the linear region of the dose-response curves for the two organisms giving moderate and readily measurable levels of inactivation (generally between 0.5 $\log_{10}$ and 2.0 $\log_{10}$). Wavelengths in the 250 nm to 270 nm range appeared to be most effective for *C. parvum* oocyst inactivation, producing approximately a 2-$\log_{10}$ reduction in infectivity. At the extremes of 216 nm and 291 nm, inactivation was lower. For MS-2, wavelengths centered in the 261 nm to 271 nm range resulted in approximately 2-$\log_{10}$ reduction and was reduced by 0.5 $\log_{10}$ to 1 $\log_{10}$ reduction for radiation from the 255-nm filter. In addition, the exposure in the 216-nm range produced a $\log_{10}$ inactivation of >3 in one experiment. UV radiation in the 230 nm and 280 nm to 290 nm range also produced comparatively inactivation.

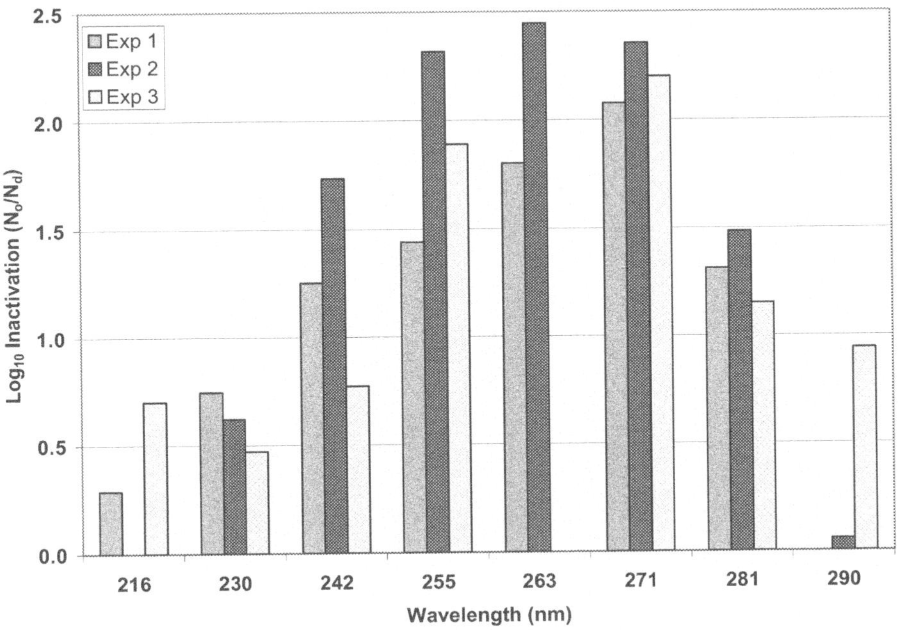

Figure 2-10. Relative UV Wavelength Effectiveness for *C. parvum* Oocysts (at 2 mJ/cm²).

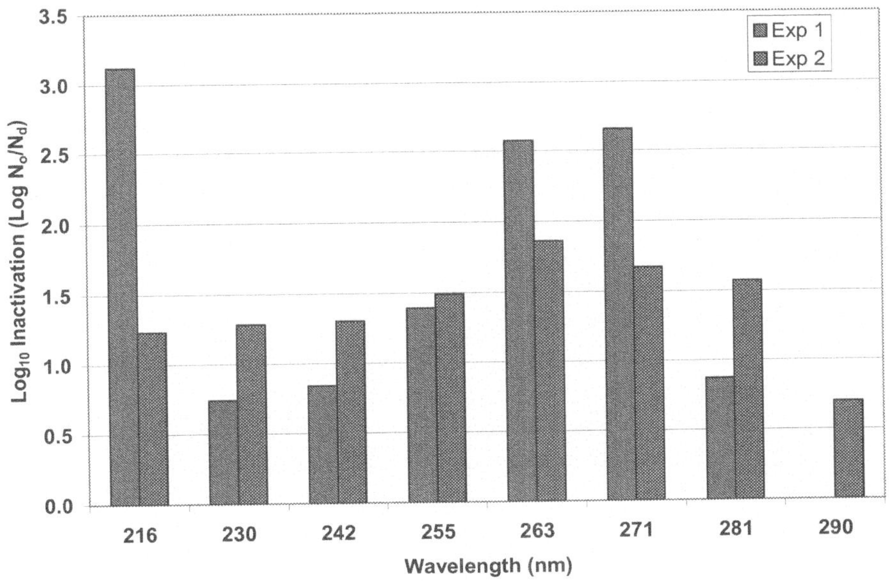

Figure 2-11. Relative UV Wavelength Effectiveness for Coliphage MS2 (at 30 mJ/cm²).

To assess the relative importance of the 254-nm range as compared to the rest of the germicidal UV spectra for inactivation of *C. parvum* oocysts and coliphage MS2, the average results were plotted relative to inactivation using the 255 nm filter, with inactivation for this filter experiment set as a relative value of 1.0, as illustrated in Figure 2-12.

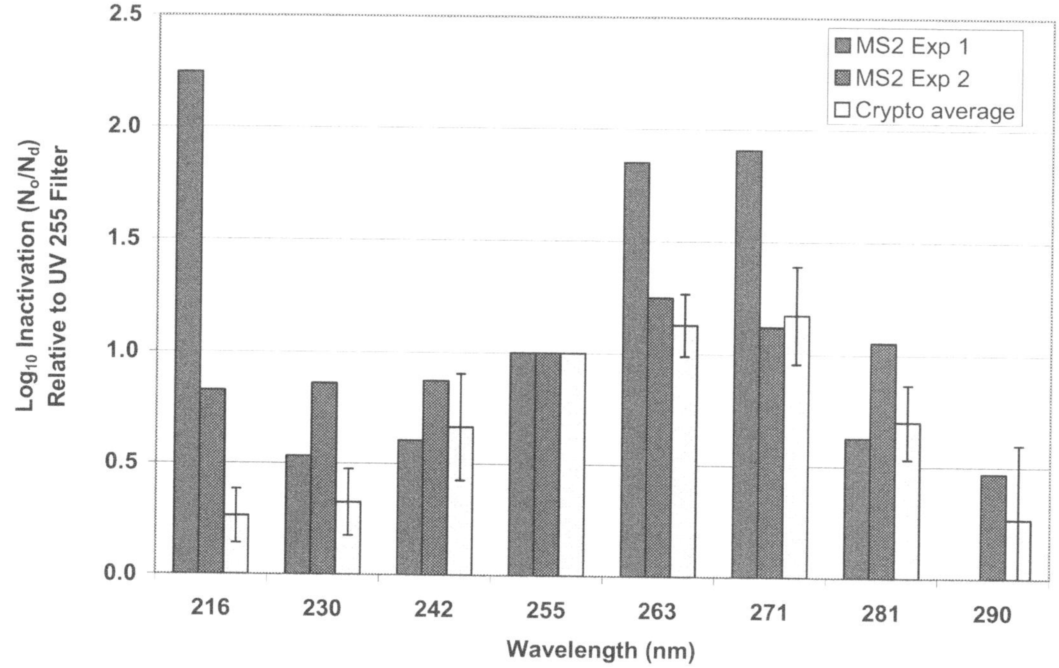

Figure 2-12. Comparative Inactivation of Coliphage MS-2 and *C. parvum* oocysts at Different Germicidal UV Wavelengths, Relative to UV 255 Filter. Error Bars Represent Standard Deviations for 2 or 3 Replicate Experiments.

The wavelengths of peak effectiveness for *C. parvum* oocyst inactivation were centered in the 250 nm to 270 nm range. The relative effectiveness of UV wavelengths at 260 nm to 270 nm is similar to that in the 254 nm range. Thus, there appears to be no particular wavelength-based advantage in the use of polychromatic over monochromatic UV systems for *Cryptosporidium* oocyst inactivation. The MS-2 results were a bit more variable between the two experiments. However, they indicate that compared to irradiation in the 254 nm range, UV at 260 nm to 270 nm and <220 nm may be more efficacious for virus inactivation compared to the 254 nm wavelength. Compared with *Cryptosporidium*, the MS2 is more susceptible to wavelengths <240 nm relative to UV 254. Therefore, polychromatic systems may have an inherent advantage over monochromatic UV-254 nm systems if significant energy can be transmitted to the organism at wavelengths near 216 nm and 265 nm. Due to the very high absorbance of wastewater below wavelengths of 240 nm from the presence of nitrate and organic matter, it is unlikely that these wavelengths would be adding to the disinfection action. Nonetheless, these results indicate a need for further investigation of wavelength effectiveness for other viruses.

2.5 Reactivation Experiments for *C. parvum* Oocysts

2.5.1 Introduction and Purpose

Many organisms do have the ability to repair DNA lesions caused by exogenous (e.g., UV light, ionizing radiation, environmental chemicals) and/or endogenous (e.g., oxidative damage, structural instability) mechanisms (Friedberg et al., 1995). Some of the health-related microorganisms do have one or more of these repair pathways, such as most of the indicator bacteria (Whitby et al., 1984; Harris et al., 1987), and some pathogenic bacteria (U.S. Environmental Protection Agency [EPA], 1986; Das et al., 1981). The level of DNA repair varies depending on the type and state of microorganism, initial UV dosage applied, and different repair conditions, but sometimes it reaches several orders of magnitude, especially when the initial UV dosages are low (Harris et al., 1987; Zimmer and Slawson, 2002). There is also a certain UV dosage for a given microorganism beyond which little or no repair of UV damage is observed (a threshold dosage) (Hoyer, 1998). Therefore, it is crucial to identify the required dosages (threshold dosages) of UV irradiation that ensure low or no DNA repair of UV-irradiated microorganisms and thereby optimize the effective doses of UV irradiation to inactivate various waterborne pathogens to acceptable levels. The UV doses required to achieve appreciable inactivation of *C. parvum* oocysts are relatively low. However, it remains necessary to characterize the extent and kinetics of repair in UV-irradiated *C. parvum* oocysts. The next step is to identify the threshold UV dosage for this microorganism in order to reduce the quantity present in drinking water to an acceptable level. Therefore, in addition to developing the UV dose-response relationship for *C. parvum* oocysts exposed to UV radiation, the project team evaluated the potential for *C. parvum* oocysts to repair UV-induced DNA damage through light and dark repair pathways at different UV radiation doses.

2.5.2 Materials and Methods

The LP UV system, UV dosimetry, UV disinfection experimental protocol, and infectivity assay are described above.

2.5.2.1 DNA Repair Experiments

Five samples, each with ~5 X 10^6 *C. parvum* oocysts in 5 mL of PBS, in 60 X 15 mm cell culture petri dishes, were exposed to low doses of UV irradiation (1.2 mJ/cm^2 or 3 mJ/cm^2) and wrapped with aluminum foil immediately after UV exposure. The experimental control was kept at 4°C. Duplicate samples were transferred to 25°C and 37°C incubators. One dish at each temperature was illuminated by a 15-watt fluorescent lamp at a distance of 25 cm with slow stirring (light repair), and the other was just stirred while wrapped with aluminum foil (dark repair). Conditions were 25°C for 2 to 4 hours and 37°C for 1 to 2 hours, as commonly used in bacterial and mammalian cell DNA repair of UV damage (8, 10, 25). These repair assay protocols were shown to be able to elicit from 0.25 log to up to 1 log recovery of *E. coli* B, depending on the conditions tested. After incubation, the samples were immediately serially diluted and inoculated into cell cultures for infectivity assay.

2.5.3 Results

Table 2-2 summarizes the results of experiments investigating dark and light DNA repair of UV-irradiated *C. parvum* oocysts dosed with 1.2 mJ/cm^2 and 3 mJ/cm^2 of LP UV. The reductions of *C. parvum* oocyst infectivity by both doses of LP UV were not restored by exposing UV-irradiated oocysts to either light or dark repair conditions. At a UV dose of 1.2 mJ/cm^2, the $\log_{10}$ reductions of *C. parvum* oocyst infectivity for control, dark repair, and light repair samples were not significantly different at the 5% level by one way or repeated measures analysis of variance (ANOVA). At a UV dose of 3 mJ/cm^2 *C. parvum* inactivation was >2.6 $\log_{10}$ and there was no restoration of this infectivity reduction by exposure of the oocysts to light or dark repair conditions. Therefore, there was no evidence of either light or dark repair of DNA damage caused by LP UV irradiation of *C. parvum* oocysts at doses causing relatively low

(~1 $\log_{10}$) and high (>2.6 $\log_{10}$) infectivity reductions and associated nucleic acid damage. These results were later confirmed by other research groups even with a different cell line (HCT-8), at a different temperature (5 °C), and with longer incubation times (up to 5 days) (Oguma et al., 2002, Zimmer et al., 2003).

Table 2-2. Effect of Dark and Light Repair Conditions on Infectivity of *C. parvum* Oocysts Dosed with 1.2 mJ/cm² and 3 mJ/cm² of LP-UV *.

Log10 *C. parvum* reduction			
Conditions	**Control**	**Dark**	**Light**
1.2 mJ/cm²			
37 °C, 1 hour	0.8	0.8	0.9
37 °C, 2 hour	0.8	1.7	1.0
25 °C, 2 hour	0.8	0.9	1.0
25 °C, 4 hour	0.7	1.4	0.9
3 mJ/cm²			
37 °C, 1 hour	>2.6	>2.6	>2.6
25 °C, 2 hour	>2.6	>2.6	>2.6

*Average values of duplicate or triplicate experiments.

CHAPTER 3.0

EFFECT OF DISINFECTANTS ON INDUCTION OF ENTEROBACTERIA INTO THE VBNC STATE

3.1 Introduction

Bacteria in the VBNC state fail to grow on the routine bacteriological media on which they would normally grow and develop into colonies, but they remain alive and capable of renewed metabolic activity (Oliver, 2000b). Cells in the VBNC state appear to be fairly inactive metabolically, but are capable of "resuscitation" to an actively metabolizing and infectious state. Such cells differ from "injured" bacteria in that the latter demonstrate an inability to grow on selective media (Mackey, 2000; McFeters and LeChevalier, 2000). In contrast "viable but nonculturable" cells show little growth on any medium, even when non-selective.

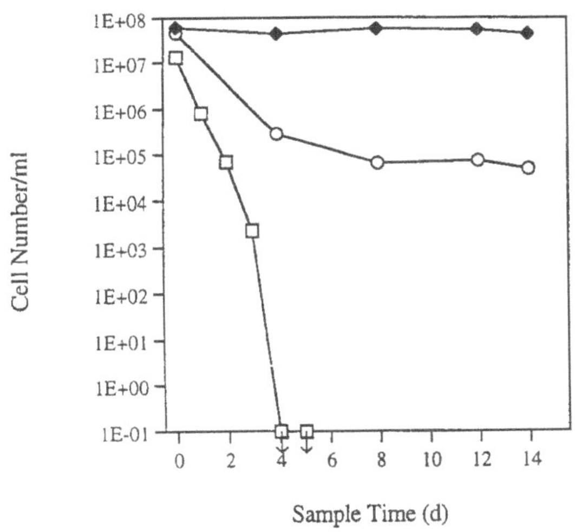

Figure 3-1. VBNC response of *V. vulnificus* upon incubation at 5°C. Shown: Total Bacterial Counts (♦), DVCs (○), and Plate (Culturable) Counts (□).

The classic VBNC response is shown in Fig. 3-1, which illustrates entry of the human pathogen, *Vibrio vulnificus*, into the VBNC state when incubated at 5°C. As demonstrated, exposure to one or more environmental stresses results in a regular decline in colony forming units (cfu; "standard plate counts"). At the same time, "total cell counts" (which do not indicate cell viability but simply the presence of a cell), typically remain constant. The critical factor that describes VBNC cells is the "viability" assay. A variety of such assays are employed in different laboratories to demonstrate this parameter (see below), but in all instances are designed to provide an estimate of the number of alive cells the population, regardless of their ability to develop into colonies on selective or nonselective media.

3.1.1 Inducers of the VBNC State

Unlike conditions that induce injury (typically agents such as antibiotics, chlorine, and other xenobiotic chemicals; Oliver, 1993), cells enter the VBNC state in response to one or more natural stresses, including nutrient deprivation, temperature shifts, elevated osmotic concentrations (e.g., seawater or food brine), oxygen concentration, and exposure to white light (Oliver, 2000c). In all cases, inducers appear to be environmental factors that are potentially injurious or lethal to a given bacterial species. The rapidity of the VBNC response to such stresses often depends on the specific bacterium, and on the physiological age of the culture (Oliver et al., 1991), with nonculturability times of hours to months reported. However, significant variation in time to nonculturability occurs between species, and even strains, of a given genus.

3.1.2 Methods for Determining Viability

A total cell count is often obtained through the use of 4',6-diamindino-2-phenylindole, dihydrochloride (DAPI) or acridine orange staining. However, viable cell counts are generally determined by the substrate responsive assay of Kogure et al. (1979), hydrolysis of 5-cyano-2,3-di-4-tolyl-tetrazolium chloride (CTC) or reduction of 2-(*p*-iodophenyl)-3-(*p*-nitrophenyl)-5-phenyltetrazolium chloride as an indication of metabolic activity or determination of cytoplasmic membrane integrity (*Bac*Light® assay). These methods have been described in detail in several reviews (Oliver, 1993, 2000b; McFeters et al., 1995; Breeuwer and Abee, 2000; Créach et al., 2003). Direct viable count (DVC) methods (e.g., Kogure et al. [1979]) are generally useful for all gram-negative bacteria, whereas Regnault et al. (2000) concluded that such methods are not routinely applicable to the various gram-positive bacteria that occur in complex ecosystems. In contrast, the *Bac*Light® method appears to be evolving as a more universal method and is the method most commonly employed by investigators studying the VBNC state.

3.1.3 Bacteria Entering the VBNC State

The number of species described to enter the VBNC state constantly increases, with nearly 60 now reported to demonstrate this physiological response. Included are a large number of pathogens, including enterohemorrhagic *E. coli*, *Salmonella* and *Shigella* spp., *Listeria monocytogenes*, and numerous pathogenic vibrios.

3.1.4 Cell Biology of the VBNC State

Cells entering the VBNC state typically undergo dwarfing, possibly a result of reductive division, although this has not been investigated. There are major metabolic changes that occur, including reductions in nutrient transport, respiration rates, and macromolecular (protein, DNA, RNA) synthesis (Porter et al., 1995; Oliver, 2000a). However, novel starvation and cold shock proteins have been observed to form during this time (Morton and Oliver, 1994; McGovern and Oliver, 1995). Adenin tri-phosphate (ATP) levels, generally observed to decline rapidly in dead and moribund cells, have generally been found to remain high in VBNC cells (Beumer et al., 1992; Federighi et al., 1998). Recent studies have demonstrated the continued expression of selected genes by cells in the VBNC state (Lleò et al., 2000, 2001; Yaron and Matthews, 2002).

Extensive modifications in cytoplasmic membrane fatty acid composition have been reported (Day and Oliver, 1999), a response likely necessary for the continued membrane potential that has been reported by some investigators (Porter et al., 1995; Tholozan et al., 1999). Similarly, significant modifications in cell wall peptidoglycan have been reported (Signoretto et al., 2000). Plasmids appear to be retained (Porter et al., 1995), and their presence or absence (in several Pseudomonas species) has dramatic consequences on whether a given factor will induce cells into the VBNC state (McDougald et al., 1995). Genomic changes have also been suggested (Bej et al., 1997; Warner and Oliver, 1998).

3.1.5 Evidence for the VBNC State

A few investigators (e.g., Kell et al., 1998; Barer and Harwood, 1999; Begosian and Bourneuf, 2001) have argued against the existence of a VBNC state in bacteria, although much of their concern addresses terminology. However, the overwhelming body of literature, which covers a variety of bacterial species, argues that this is a novel physiological state induced by a variety of environmental factors. Indeed, some investigators who have previously argued against the VBNC state now offer evidence for it (Mukamolova et al., 1998a, 1998b, 1999) and/or have confirmed what has been previously shown by others (e.g., Begosian et al., 2000). A few of the studies that offer compelling evidence for the VBNC state are noted below.

Using multi-parameter flow cytometry, Porter et al. (1995) studied *E. coli* cells as they entered the VBNC state in filter-sterilized lake water. Measurements of membrane potential, membrane integrity, and intracellular enzymatic activity "...provided extensive evidence for the validity of the methods for monitoring cell viability during adoption of the viable-but-non-culturable state in starved *E. coli*." Beumer et al. (1992) determined that the levels of ATP remained fairly constant within cells of *Campylobacter jejuni* following incubation in saline at 20°C, conditions that led to complete nonculturability.

Yaron and Matthews (2002) found that a variety of genes, including *mobA, rfbE, stx1*, and those for 16S rRNA synthesis, were expressed in nonculturable cells of *E. coli* O157:H7. Similarly, Barrett (1998) and Saux et al. (2002) reported continued production of message for several genes after cells of *V. vulnificus* were in the VBNC state for as long as 4.5 months. Such findings strongly indicate not only viability, but active involvement in the VBNC process. Similarly, Rahman et al. (1994) found that *Shigella dysenteriae* cells that had been in the VBNC state for 4 to 8 weeks were capable of active uptake of methionine and its incorporation into protein.

Significant, and what might be characteristic biochemical changes in VBNC cells, have also been documented. Signoretto et al. (2002), in studying the cell wall peptidoglycan of *E. coli* cells entering the VBNC state, reported an overall increase in cross-linking, a three-fold increase in unusual DAP-DAP cross-linking, an increase in muropeptides bearing covalently bound lipoprotein, and a shortening of the average length of glycan strands in comparison to exponentially growing cells. VBNC cells were also found to have an autolytic capability far higher than that measured in exponentially growing cells. Similar findings were reported by Signoretto et al. (2000) for *Enterococcus faecalis*.

3.1.6 Detection of Cells in the VBNC State

Cells in the VBNC state are, of course, nonculturable by routine bacteriological methods. Some studies have observed prolonged culturability when certain anti-reactive oxygen species (e.g. catalase) are included in the culture medium (see Section 3.1.8). While this observation has improved understanding of the VBNC response in some bacteria, the inclusion of such compounds has not prolonged culturability in many bacteria.

The most obvious method of detecting dormant cells would appear to be the use of a molecular method, such as polymerase chain reaction (PCR), which does not require culture of the organism in question. However, PCR technology may not be effective for this purpose. Using PCR to amplify the hemolysin gene of *V. vulnificus*, approximately 500 times more DNA from cells in the VBNC state was required for signal detection than for cells in the culturable state (Brauns et al., 1991; Brauns and Oliver, 1994). This was true whether whole cell lysates or extracted DNA was employed for gene amplification. Subsequently, Coleman and Oliver (1996) compared the conditions for amplification of this gene, and again found that at least 100 times more cells were required from the VBNC state, compared to the culturable state. In more recent studies, Warner and Oliver (1998) employed the more sensitive randomly amplified polymorphic (RAPD) PCR method to monitor changes in the DNA of cells of *V. vulnificus* as these cells entered the VBNC state. Banding changes (suggestive of genomic rearrangements) and a gradual loss of amplification was reported, with total loss typically occurring within 7 days. Such changes could be reversed by subjecting the cells to a temperature upshift prior to PCR amplification, resulting in their resuscitation. The same observation has been reported by Bej et al. (1997). Thus, this procedure does not hold great promise as a method for detecting bacteria in air.

3.1.7 Resuscitation from the VBNC State

Sides et al. (1999) reported cells of *V. vulnificus* survive in the VBNC state for as long as 5 years. Ultimately, however, for the VBNC state to be of consequence, the cells must be able to respond to some environmental stimulus, become metabolically active and as a consequence, culturable. This process is generally referred to as "resuscitation."

The simplest form of resuscitation is observed following removal or reversal of the factor(s) that induced nonculturability. The best-studied organism is *V. vulnificus*. This aquatic pathogen is induced into the VBNC state by incubation at temperatures below 10°C (Wolf and Oliver, 1992). This occurs regardless of nutrient environment (Oliver and Wanucha, 1989) or under starvation conditions (Oliver et al., 1991). A simple temperature upshift results in resuscitation of these cells. This has been demonstrated *in vivo* (Oliver and Bockian, 1995), *in situ* (Oliver et al., 1995) and *in vitro* (Wolf and Oliver, 1992; Oliver et al., 1991).

While many bacteria have been reported to enter the VBNC state, fewer have been conclusively demonstrated to return to the actively metabolizing and culturable state. While bacteria like *V. vulnificus* are easily resuscitated by a simple reversal of the inducing factor, the difficulty in demonstrating resuscitation was exemplified by Steinert et al. (1997). They examined nonculturability and resuscitation in *Legionella pneumophila*. Entry into this state was easily induced by nutrient starvation, but resuscitation could only be demonstrated following co-incubation of the VBNC cells with the amoeba, *Acanthameoba castellani*. Thus, resuscitation may prove to be the most difficult aspect of studying the VBNC state in some bacteria.

Other "natural" resuscitation conditions have been reported by Pommepuy et al. (1996), who studied the induction of *E. coli* into the VBNC state by solar irradiation. Using quartz chambers and *in situ* incubation, they showed natural cycles of culturability and nonculturability in lake water that correlated with the levels of irradiation. Visible light has also been shown to induce dormancy in *E. coli* in natural river water (Barcina et al., 1989).

3.1.8 Culture of VBNC Cells

In an important advancement in the development of methods to allow culture of "nonculturable" cells, Mizunoe et al. (1999) determined that enterohemorrhagic *E.* coli cells that were nonculturable (<10^0 cfu/ml) on agar media following incubation in distilled water at 4°C could be cultured to a high level (10^4 cfu/mL to 10^5 cfu/mL) when plated on the same media containing either catalase, sodium pyruvate, or α-ketoglutarate. A similar finding was reported (Sides et al., 1999) for O157:H7 and non-O157:H7 cells of *E. coli*. This suggests that the metabolic by-product H_2O_2 is produced by cells during the VBNC process, or is present in culture media, but that this toxic agent could be neutralized by these anti-reactive oxygen species compounds.

Mizunoe et al. (2000) subsequently reported *V. parahaemolyticus* enters the VBNC state in ca. 12 days following starvation and 4°C incubation. However, when the plating medium was supplemented with catalase or sodium pyruvate, culturability was maintained for a considerably longer period. Similarly, Wai et al. (1996) reported that *A. hydrophila* cells enter the VBNC state in ca. 45 days when incubated under starvation conditions at 4°C. Cells could be cultured, however, when either of these two agents was present in the plating medium. Similar results were reported for *V. vulnificus* by Sides et al. (1999) and Begosian et al. (2000).

Whitesides and Oliver (1997) are believed to be the first to suggest that the plating medium may be important in the VBNC state. They stated that "...elevated nutrient might be toxic in some manner to cells in this [the VBNC] state". This hypothesis was testable, and several studies (Begosian et al., 2000; Vedeikis and Oliver, 2000; Mizunoe et al., 1999, 2000) indicated that the presence of reactive oxygen species (ROS) in solid media, or produced by VBNC cells when plated to solid media, might prevent the culture of cells present in this state. The value of such ROS scavengers is indicated in Figure 3-2, which demonstrates the effect of incorporation of catalase or pyruvate in plated media (HI agar) onto which VBNC cells of *V. vulnificus* are placed.

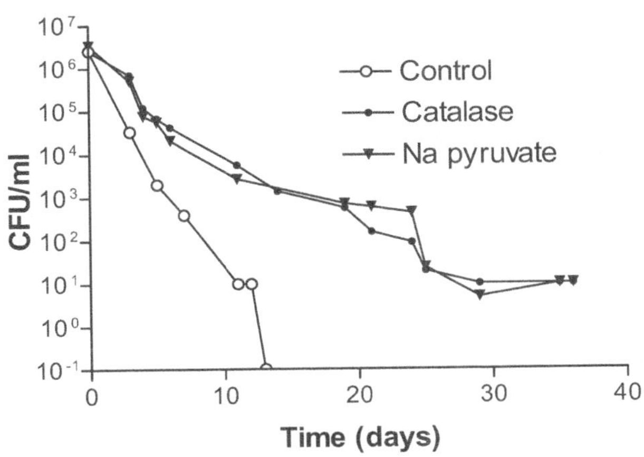

Figure 3-2. The Effect of Incorporation of Radical Scavengers Catalase or Pyruvate into Plated Media (HI agar) During Monitoring of VBNC Cells of *V. vulnificus*

Barer (1997) has suggested that, since viability (as classically defined) can only be demonstrated by colony development, the term "viable but nonculturable" is an oxymoron. Further, given the finding that some cells, when present in the VBNC state, may be at least partially culturable under certain conditions (e.g., the presence of ROS-scavengers), Oliver (2000b) also suggested that this phrase is not fully accurate. However, the phrase "viable but nonculturable" has become adopted by the majority of researchers investigating this phenomenon. Certain conditions do induce these cells to lose their ability to develop colonies on routine media, and to thus become "nonculturable" by routinely employed methods. Further, the morphological, physiological, biochemical, and genetic changes such cells undergo are dramatic, and may also prevent their detection by more sophisticated molecular (e.g., PCR) and/or spectral analyses.

3.1.9 Virulence of Cells in the VBNC State

A key aspect of the VBNC state is whether cells in the VBNC state retain virulence. Such cells generally demonstrate extremely low metabolic rates, and with few exceptions, cells in such a near-dormancy state might not be expected to be able to initiate infection. Note, however, that VBNC cells appear to actively produce mRNA, and this could result in the continued synthesis of one or more virulence factors. For example, Saux et al. (2002) recently demonstrated that *V. vulnificus* continues to produce mRNA for its hemolysin/cytolysin even after 4.5 months in the VBNC state. Similarly, Nwoguh et al. (1995) demonstrated that non-culturable cells of *E. coli* and *Salmonella* strains retain inducible enzyme activity. Likely more important, however, would be resuscitation under the appropriate conditions that may lead to subsequent infection of a susceptible host.

To date, and depending on the bacterium in question, the evidence has often been contradictory. Certainly, numerous studies have demonstrated the ability of pathogens that were present in the VBNC state to initiate infection when inoculated into a susceptible host. These include *C. jejuni, E. faecalis, E. coli,* several *Salmonella* species, *S. dysenteriae, L. pneumophila,* and *Helicobacter pylori* (Oliver, 2000a, 2000b). As specific examples, Colwell et al. (1985) inoculated VBNC cells of *V. cholerae* into ligated rabbit ileal loops, and observed positive responses (i.e., enteropathogenicity) in all samples. Intestinal fluid was also reported to contain culturable *V. cholerae* cells. In a subsequent study, Colwell et al. (1996) reported a study wherein two human volunteers ingested VBNC cells of *V. cholerae*. Approximately 48 hours after the challenge, one of the volunteers had culturable *V. cholerae* cells in his stool, while at 5 days the second volunteer was stool-positive for this pathogen.

As a second example, Oliver and Bockian (1995) demonstrated the ability of *V. vulnificus* cells in the VBNC state to retain virulence against mice. Cells were made VBNC (<0.1 cfu/mL) by incubation at 5°C. These cells were diluted such that, while the preparation contained 10^5 cells in the VBNC state, there were <0.04 cfu/mL. Mice inoculated with this preparation died, and culturable *V. vulnificus* cells were recovered from blood and the peritoneal cavity.

3.2 Major Goals of VBNC Study

One focus of the research presented in this report was to determine whether microbes in wastewater, on irradiation with UV light or exposure to chlorine or chloramines, enter into the VBNC state. Specifically, the goals of this aspect of the study were to address the following questions:

- How do natural environmental factors affect induction of enteric bacteria (*E. coli* [F+ and F- strains], *Salmonella typhimurium* [plasmid-bearing and without plasmid], and *Aeromonas hydrophila*) into the VBNC state? These studies examined primarily temperature (5°C, 15°C, and 22°C) effects on cells in PBS and natural wastewater (sterile).

- Does UV irradiation induce cells into the VBNC state? These studies were also performed on cells present in PBS and wastewater.
- Do chlorine or chloramines induce cells into the VBNC state? These studies were also performed on cells present in PBS and wastewater.
- Can cells, induced into the VBNC state by UV irradiation or chlorination be resuscitated? If so, what are the environmental factors and times required?
- Can cells in the VBNC state be detected by PCR?

3.3 Materials and Methods

3.3.1 Bacterial Strains and Culture

E. coli K12 [isogenic F+ and F- strains], *S. typhimurium* WG49 and WG45, and *A. hydrophila* were used. Strains K12 F^+ and WG49 carry plasmids encoding for ampicillin and kanamycin resistance, respectively, whereas K12 F^- and WG45 lack these elements. Strains were routinely maintained on Luria Bertani (LB) agar plates.

3.3.2 Preparation of Microcosms

Cells were grown at 37°C to log phase in L-broth. These cells were then washed once in PBS then diluted into 40 mL microcosms of either PBS or sterile wastewater (see below) to yield ca. 10^6 cfu/mL. Wastewater was nonchlorinated, secondary-treated at pH 8.0, acquired from a wastewater treatment plant in Charlotte, North Carolina. The wastewater was sterilized by filtering through a 0.4-μm filter and then autoclaving at 121°C for 20 minutes. These microcosms were then placed at appropriate temperatures and monitored over time by performing plate counts (LB agar) to determine culturability under various conditions.

3.3.3 Cell Preparation

A variety of methods for preparing cells in different physiological states were employed as described below.

3.3.3.1 Stationary Phase Cells

One colony of each of the strains was inoculated into LB broth and grown overnight with aeration at room temperature (22°C). An inoculum (0.1 mL) of these cells was placed in 10 mL of fresh LB broth and shaken at 37°C until the stationary phase was reached (OD_{610} of ca. 0.80). Aliquots (1 mL) were centrifuged (ca. 14,000 x g for 5 minutes), the supernatant carefully removed, and the pellet resuspended in PBS. The cells were then centrifuged and washed a second time with the pellet resuspended in 1 mL PBS.

3.3.3.2 Log Phase Cells

Cells were prepared as described above except that they were grown to a final OD_{610} of 0.15 to 0.20.

3.3.3.3 Starved Cells

Cells were grown to log phase and prepared as described above but were then inoculated (1 mL) into PBS (5 mL) and incubated for 20 to 24 hours at room temperature.

3.3.3.4 Cells Grown in Wastewater

In the possibility that cells that were adapted to wastewater might be more resistant to chlorination, a single bacterial colony was inoculated into filtered, autoclaved wastewater (see above) and shaken overnight at room temperature. These cells (0.05 mL) were then inoculated into 5 mL of filtered, autoclaved wastewater and shaken for 5 days at room temperature.

3.3.3.5 Cell Growth/Survival in Wastewater

Log phase cells were inoculated (1% final concentration) into sterile wastewater. These microcosms were then placed at 5°C, 15°C, or 21°C and monitored for culturability.

3.3.4 Chlorination of Cells

A 1% dilution of commercial bleach (6% sodium hypochlorite) was made in filtered and autoclaved deionized water. This was filter-sterilized using a 0.2 μm Acrodisc® filter (Pall Corporation, Ann Harbor, Michigan). Two 50-mL wastewater microcosms were prepared in sterile, disposable centrifuge tubes. One tube served as the experimental sample while the other served as a control. Cells (0.5 mL except for 2.0 mL of wastewater-grown cells) were added to each wastewater microcosm and shaken to ensure even distribution. Sodium hypochlorite (FC) solution was then added to provide an initial concentration of 1 mg/L of FC in the wastewater. Samples were taken at times 0 (ca. 20 sec), 1 minute, 5 minutes, 10 minutes, and 60 minutes and placed into 20% sodium thiosulfate (1:1 volume/volume in autoclave buffered saline) to inactivate any residual FC. Total and FC levels in the microcosms at the initiation of the study and after the 10-minute sample time were determined using the N,N-diethyl-p-phenylenediamine (DPD) method (Hach Co., Loveland, Colorado).

3.3.5 UV Irradiation of Cells

As described for the chlorination studies, cells were grown to log phase, then washed with PBS and inoculated into sterile PBS/wastewater to provide ~10^6 cfu/mL. Cells were irradiated with LP UV light at 0, 50 mJ/cm^2, or 100 mJ/cm^2, then immediately covered with foil to prevent photoreactivation by white light. Irradiated cells were analyzed for culturability, for total cells by acridine orange direct count (AODC), and for DVC by the methods described below.

3.3.6 Cell Enumeration

Cells from control (untreated) and chlorine-treated or UV-irradiated microcosms were removed, serially diluted, and plated onto LB agar to determine the number of culturable cells present at various treatment times. The method of Kogure et al. (1979) was used to enumerate total and viable (substrate responsive) cells. Briefly, samples at 0, 10 minutes, and 60 minutes were incubated with 0.025% yeast extract (Difco, Sparks, Massachusetts) and 0.002% nalidixic acid (Sigma, St. Louis, Missouri) to inhibit the septation of cells growing in response to the nutrient addition. After 12 hours, samples were fixed with 50 μl 37% formalin and stained with filter sterilized 0.1% acridine orange. Samples were then filtered through 0.2 µm black polycarbonate filters (Fisher Scientific). A control sample lacking Kogure reagents was employed to determine spontaneous elongation, and these values were subtracted from experimental data points. Epifluorescent microscopy (Olympus BX51) using appropriate filters was used to enumerate total and viable cells. This method has a limit of detection of ca. 5×10^3 cells/ml.

3.3.7 Resuscitation Studies

To attempt to recover culturable cells from the VBNC state (whether chlorine or UV induced), ROS scavengers, a temperature upshift, and the use of media solidified with agarose instead of agar were examined as described below.

3.3.7.1 ROS Scavengers

As there is evidence that ROS may be involved in the VBNC state of some bacteria (Vedeikis and Oliver, 2000; Kong et al., submitted 2004), 200 units of bovine liver catalase, 600 units of superoxide dismutase, or a 1% solution of sodium pyruvate was spread onto LB agar plates which were then dried prior to use. All stock solutions (Sigma, St. Louis, Missouri) were made with filtered, autoclaved, deionized water (DI) and kept in the dark at 5°C. At each time point, 100 µl of the chlorine-exposed cells were spread onto the ROS scavenger-treated LB plates.

3.3.7.1 Temperature Upshift

Gupte and Joseph (2001) provided evidence for the resuscitation of *Salmonella enterica* cells from the VBNC state following heat shock treatment. Employing their protocol, cells (250 µl) taken at each of the sample points were exposed to 80°C for 15 seconds. These heat-shocked samples (100 µl) were plated onto LB agar.

Vibrio vulnificus cells in the VBNC state show significant culturability when agarose is substituted for agar as a solidifying agent (Kong et al., submitted 2004). LB medium was prepared with 15 g of agarose instead of 15 g of agar. At each time point, 100 µl of the chlorine-exposed cells were plated onto this medium.

3.3.8 Statistical Analysis

All experiments were conducted at least twice with each of the bacterial strains. Results were graphed and the standard error of the mean calculated and shown in the figures.

3.4 Results

3.4.1 Effect of Temperature on Induction into the VBNC State

Results can be summarized by several representative graphs of data obtained during these studies. The effects of various temperatures (5°C, 15°C, and 22°C) for inducing the VBNC state in the three bacterial species were examined. Figure 3-3 is illustrative of the results observed. Low temperature generally induced nonculturability in the various strains, while higher temperatures (e.g., >15°C) allowed long-term survival of the cells.

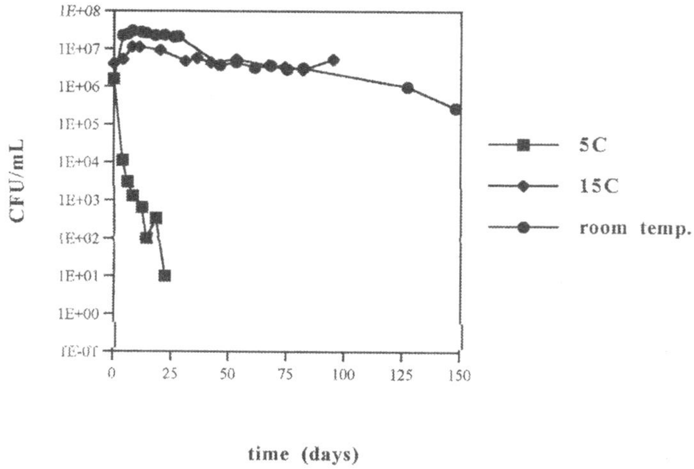

Figure 3-3. Effect of Temperature on Induction of Bacteria into the VBNC State. *S. typhimurium* WG49 in Sterile Wastewater.

3.4.2 Effect of UV Irradiation on Induction into the VBNC State

Cells of *S. typhimurium* (both strains WG45 and WG49) were grown to the log phase in Luria broth at 37°C, washed, and diluted in PBS to provide ca. 10^6 cfu/mL. Then, 40 mL were placed into petri dishes and irradiated with LP UV light at 0, 50 mJ/cm^2, or 100 mJ/cm^2. Results for the two strains are shown in Figure 3-4, which represents an average of three replicates of this study.

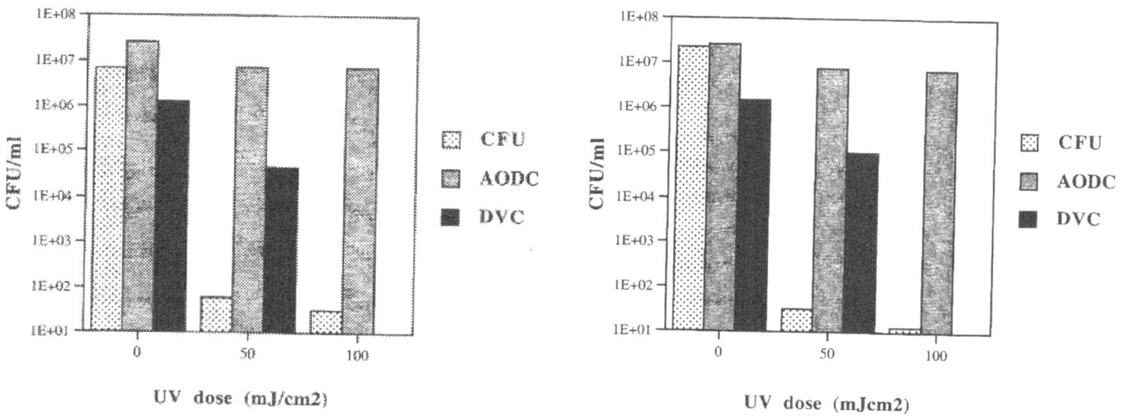

WG 45WG 49

Figure 3-4. *S. typhimurium* WG45 (left) and WG49 (right) Exposed to LP UV Radiation.

The results were remarkably consistent between strains WG45 and WG49. Levels of culturable cells decreased from ca. 10^7 cfu/mL with no irradiation to ca. 10^2 cfu/mL after 50 mJ/cm^2, and to only slightly less than this value at 100 mJ/cm^2. This represents a >99.99% to 99.999% decrease in culturability following irradiation. However, even at the highest level of UV irradiation, a small percentage of cells of both strains remained culturable. Strain WG49 appeared to be slightly more sensitive to all levels of irradiation, which might be due to its carrying the high molecular weight F$^+$ plasmid. Such plasmids are known to generally impart a metabolic load on the cells, which in this case, might make the cells more susceptible to the UV irradiation.

Despite these dramatic declines in culturability (to near sterility in some cases), AODC values indicated that the number of total cells of both strains declined very little, even at 100 mJ/cm^2. Most significantly, even when only very small numbers of cells remained culturable (at the $50mJ/cm^2$ dosage), the substrate responsiveness assay (DVC of Fig. 3-4) indicated that large (10^4/mL to 10^5/mL) viable populations remained. While this represents <1% of the total cell population, the number represented is remarkable. No DVC ("viable") cell populations were detected at the 100 mJ/cm^2 dosage, even when small numbers of culturable cells were evident. This is explainable in that the level of detection of living cells by the culture method is considerably more sensitive than that (microscopy) employed in the DVC method. Thus, at the extremely low levels of cells detectable at the highest UV dosage, finding living cells by either assay (culturability or DVC) is significant.

Because it was possible that irradiated cells might be capable of repairing their UV-induced damage by some "dark repair" mechanism, and subsequently become culturable, or that cells might not respond to our initial viability assays because of some temporary "shock" or other factors, we repeated our assays after the cells had been maintained refrigerated for 4 days. As seen in Figure 3-5, little difference was observed in the ability of these cells to grow on Luria agar (i.e., they remained nonculturable), in total cell counts (i.e., the cells were not lysing), or in substrate responsiveness (i.e., a large number of the cells appear to truly be in a VBNC state). The only exception to this was WG45, which at a UV dosage of 50 mJ/cm^2 appeared to lose almost all viability on continued incubation. The slight decrease in culturability observed in each of these experiments was expected, as the cells at 5°C are known to undergo a slow decrease at this temperature.

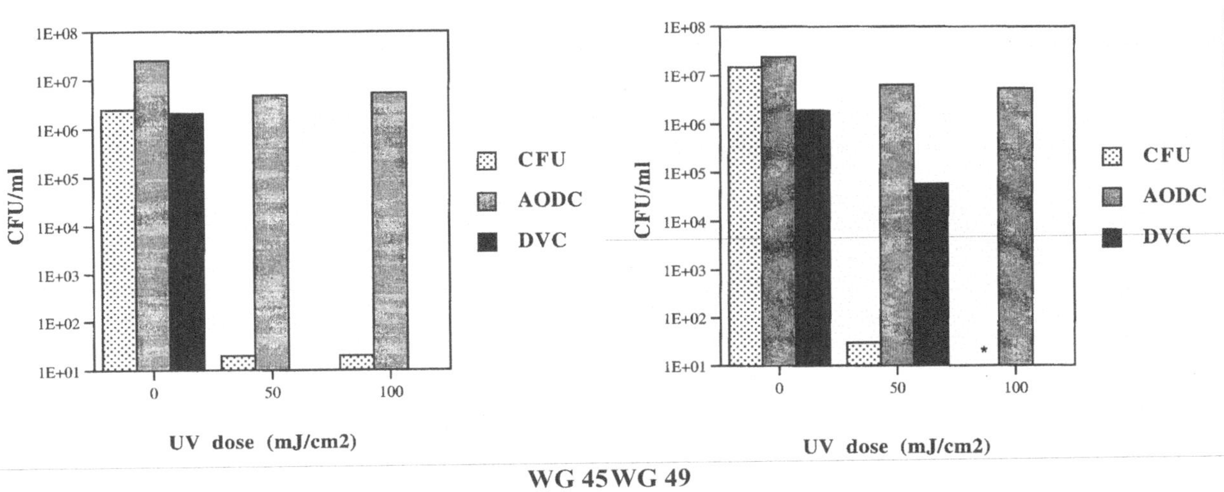

Figure 3-5. ***S. typhimurium*** **WG45 (left) WG 49 (right) 4 Days After Exposure to LP UV Radiation.**

The ability of UV irradiation to induce the VBNC state in the various enterics was investigated. Figure 3-6 illustrates the typical results obtained, indicating that increasing doses of LP UV light resulted in loss of culturability of the cells, as would be expected. The results also suggested, however, that the cells remained viable at least for 4 days following irradiation.

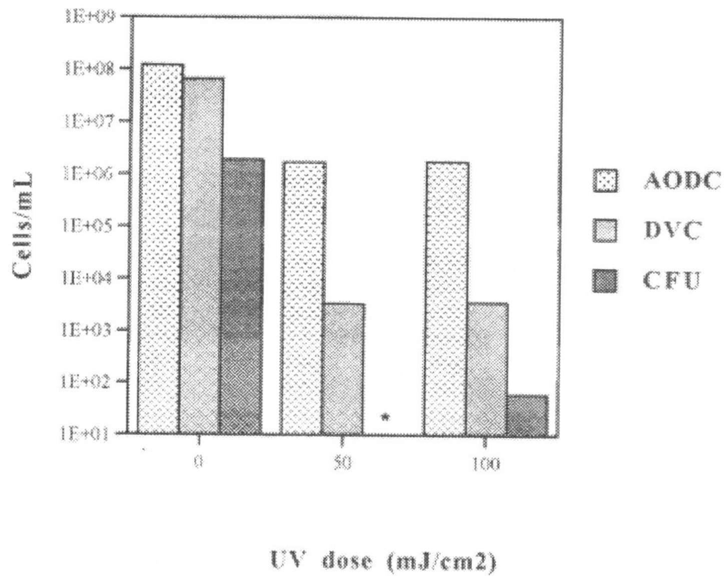

Figure 3-6. Effect of UV Irradiation on Culturability of *E. coli* K12 F$^+$, 4 Days After UV Exposure. *Indicates <10cfu/mL

Resuscitation of UV-irradiated cells was attempted by exposing the treated cells to fluorescent lights for 4 days. Figure 3-6 shows the results from one such study. Replicate studies suggest that some resuscitation of the irradiated cells of this strain may be occurring (based on increases in DVC levels after 4 days).

For the next series of studies, the same bacteria were examined for sensitivity to UV irradiation when present in sterile wastewater. Figure 3-7 shows UV induction of *E. coli* K12 F$^+$ into the VBNC state in wastewater. While UV again caused significant decreases in culturability of this strain, the extent of decrease was considerably less than that which occurred in PBS. Indeed, viability of this strain exceeded 10^5 even at the highest UV dosage. It is our belief that wastewater is protective to the cells in this respect.

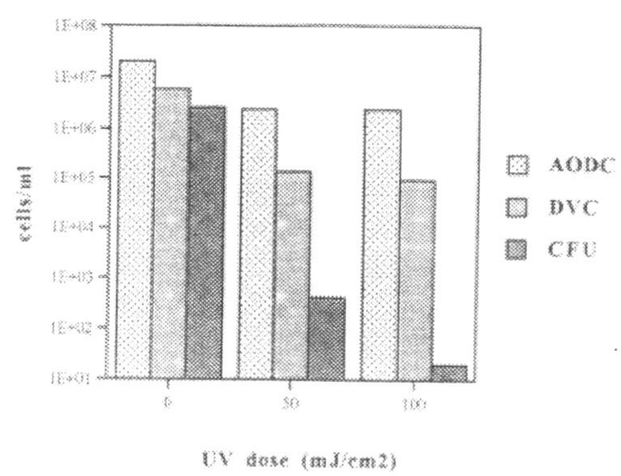

Figure 3-7. Induction of *E. coli* K12 F$^+$ into the VBNC State in Wastewater Following UV Irradiation.

As observed with the F$^+$ strain, UV induction of *E. coli* K12 F$^-$ into the VBNC state in wastewater (data not shown) revealed that considerable viability was maintained when the cells were irradiated in wastewater as compared to PBS. Both culturability and viability were almost identical for the F$^+$ and F$^-$ strains.

As observed with the *E. coli* strains, UV induction of *S. typhimurium* WG45 and WG49 (Figure 3-8) into the VBNC state in wastewater exhibited considerable survival, even at the highest UV dosages. Little difference in results was seen when comparing this response in *E. coli* and *S. typhimurium*.

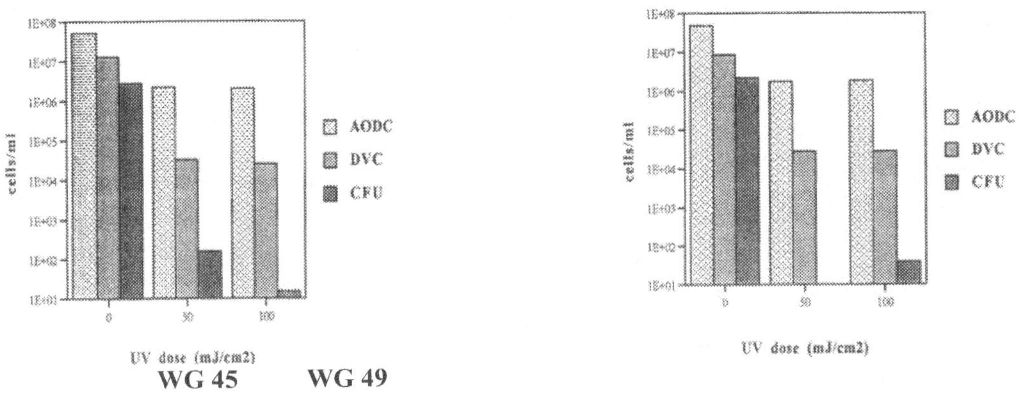

Figure 3-8. Induction of *S. typhimurium* WG45 (left) and WG49 (right) into the VBNC State in Wastewater Following UV Irradiation.

3.4.3 Effect of Chlorine/Chloramine on Induction into the VBNC State

The studies were then extended to examine whether the cells, despite being nonculturable, remained viable in wastewater following chlorination and resulting (due to FC combining with ammonia in the wastewater) chloramine treatment. This form of chlorination was found to be bactericidal, and in no case was viability maintained. Total loss of culturability ("cfu" in Figure 3-9, c-cfu is the culturability of the control sample) and viability (DVC in Figure 3-9) was observed prior to the first (1 minute) sampling (detection limit for DVC is at 10^4 cells/mL).

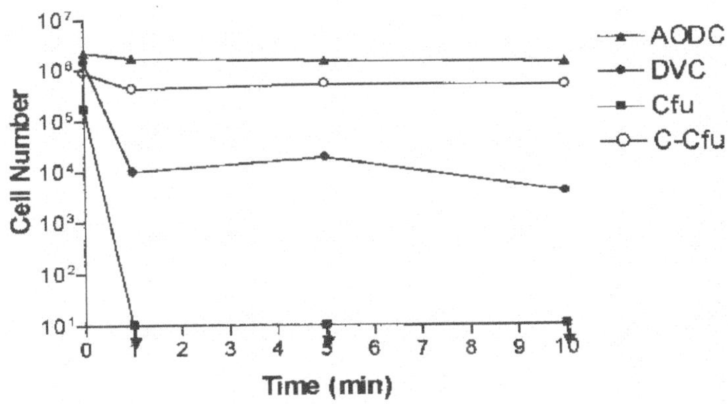

Figure 3-9. Effect of Chlorine/Chloramine on Induction of Bacteria into the VBNC State. *E. coli* K12 F⁻.

Subsequently, the cells were treated to allow production of "stress proteins," which are known to protect cells against various environmental stresses. Stress proteins were induced by growth to stationary phase and by nutrient starvation. Like unstressed cells, these also died rapidly when exposed to 1.0 mg/L chloramines in natural wastewater.

3.4.4 Resuscitation of Bacteria Induced into the VBNC State

Following the VBNC induction studies, an attempt was made to "resuscitate" cells from the viable but nonculturable state, whether induced by UV irradiation or chlorination. In the case of UV irradiated cells, resuscitation attempts were a result of allowing the exposed cells to remain in the dark in PBS, or exposing them to fluorescent light. A variety of resuscitation conditions were examined for chlorine-treated cells, including the presence of compounds that would neutralize the ROS which are known to be produced by some cells in response to VBNC conditions, and which cause loss of culturability. These included catalase, sodium pyruvate, and superoxide dismutase. Other methods, such as substituting agarose for agar, and a heat shock method shown to allow resuscitation of certain bacteria, were also tried. As shown in Table 3-1, in no case in the resuscitation studies were culturable cells ever observed. These results can be interpreted to mean that cells in the VBNC state are of no consequence, as they do not resuscitate to the actively metabolizing and potentially infectious state. However, it must be realized that the conditions required for resuscitation may not have been achieved, and that resuscitation could potentially occur in wastewater facilities or in subsequent waterlines or individual persons consuming the water.

Table 3-1. Resuscitation Conditions and Response for *E.coli* K12 F$^+$.

Time (min)	cfu/mL[a]		Control		Catalase		Sodium pyruvate		Superoxide dismutase		Heat shock		Agarose	
0.5	<1	<1	2.4 x10^7	4 x 10^7	<10	<10	<10	<10	<10	<10	<10	<10	<10	<10
1	<1	<1	2.3 x 10^7	3 x 10^7	<10	<10	<10	<10	<10	<10	<10	<10	<10	<10
5	<1	<1	2 x 10^7	2 x 10^7	<10	<10	<10	<10	<10	<10	<10	<10	<10	<10
10	<1	<1	2.3 x 10^7	3 x 10^7	2	<10	<10	<10	<10	<10	<10	<10	<10	<10
60	<1	<1	2.2 x 10^7	5 x 10^7	5	<10	5	<10	<10	<10	<10	<10	<10	<10

[a]Culturability (in cfu/mL) for two trials are shown

3.4.5 PCR Amplification of Cells in the VBNC State

Previous studies confirm that PCR can be used to amplify cells from the VBNC state. In some instances the conditions employed to this end must be optimized for VBNC cells (Coleman and Oliver, 1996). However, a major problem exists in the inability to determine whether the

DNA amplified is from VBNC cells or dead cells. Thus, while we are able to demonstrate the presence of DNA from the desired bacteria, it is not possible at this time to know conclusively that the DNA is being amplified from living cells.

3.5 Conclusions

It appears that UV irradiation induces cells to become nonculturable, but the DNA cross-linking that occurs is likely to be substantial, and little indication of significant recovery (resuscitation) from this state was evident. Chlorination also induced nonculturability. The "realistic" dynamic chloramination (chlorine/chloramine treatment resulting from addition of chlorine to ammonia containing wastewater) in wastewater was found to be highly bactericidal, with no evidence of resuscitation observed. Thus, it appears that the "real world" use of chlorine for disinfection of bacteria is largely effective and would not lead to the presence of VBNC that were capable of resuscitation.

CHAPTER 4.0

MECHANISMS OF INACTIVATION AND RESISTANCE

4.1 Introduction

Microbes differ in their susceptibility to different disinfectants. The reasons for these differences are not well known but may include morphological properties, genetic differences, nucleic acid make-up, growth patterns, tendency to aggregate, and interaction with external factors, such as water quality constituents (e.g., particles and solutes). The research described in this section was aimed at developing a better understanding of why and what differences exist among bacteria and viruses in their response to disinfectants, with a focus on UV disinfection.

4.2 UV Inactivation of Aggregated and Particle-associated Virus

4.2.1 Introduction and Purpose

Viruses in wastewater are usually aggregated and/or closely associated with cell debris and other fecal material (Sobsey, 1989). When microorganisms are aggregated and/or associated with particles, it may reduce disinfection efficacy because they are protected or shielded from exposure to disinfectants. Therefore, the extent to which viruses are associated with particulate matter in wastewater and the subsequent impact of this association on UV inactivation was determined.

4.2.2 Materials and Methods

4.2.2.1 Wastewater and its Processing

Wastewater samples were taken from Southern Durham WRF (Durham, North Carolina). Due to the low titers of coliphage in the final effluents, raw sewage (the influent to the plant) was used because it contains higher titers of coliphage. Furthermore, UV disinfection of untreated and minimally treated sewage is being proposed and is becoming more widely practiced for sources such as primary effluents, combined sewer overflows, and sewage bypasses from treatment facilities during wet weather. Therefore, the extent to which particles in poorly treated sewage and sewage-laden effluents influence the inactivation of viruses and other microbes is of interest and concern.

The raw sewage samples were processed in the laboratory into three matrices with respect to particulate content (e.g., different particle size distributions): supernatant, settled, and resuspended fractions. The supernatant sample was prepared by centrifuging the wastewater at 6700 g for 30 minutes and recovering the supernatant. The settled sample was prepared by gravity-settling raw sewage for 4 hours and recovering the supernatant. The (centrifuged and) re-suspended sample was prepared by centrifuging raw sewage at high speed, inactivating microbes in the supernatant with high UV doses, and resuspending the centrifuged particulates in the UV-irradiated supernatant. The typical size distributions of these three matrices are shown in Figure 4-1.

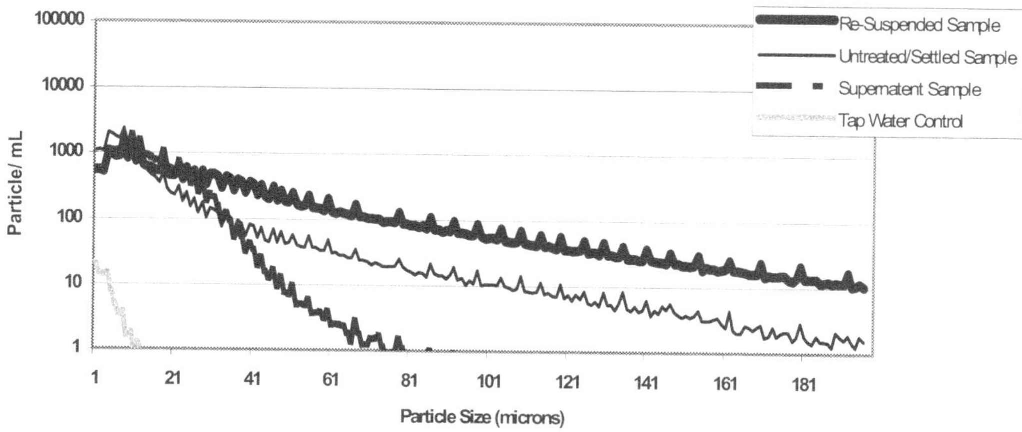

Figure 4-1. Typical Particle Size Distribution of the Three Wastewater Matrices Tested.

4.2.2.2 UV Disinfection

UV irradiation was performed as described in Section 1.1.2.5. Briefly, 5-mL volumes of each processed wastewater samples in 60 X 15 mm cell culture petri dishes were irradiated with a UV-collimated beam system while magnetically stirring the samples slowly (~30 rpm) at room temperature (23°C to 25°C). After predetermined exposure times corresponding to target UV doses, samples were removed from the UV irradiation system and diluted serially 10-fold for subsequent infectivity assays.

4.2.2.3 Microbial Assays

Somatic and male-specific coliphage were assayed as in an EPA method (Method 1602: Male-specific (F+) and somatic coliphage in water by single agar layer [SAL] procedure). Briefly, both test samples of wastewater and bacterial hosts grown overnight (CN13 for somatic and F_{amp} for male-specific coliphage) were added to 2X tryptic soy agar (TSA) that contained antibiotics (nalidixic acid for CN13 and ampicillin/streptomycin for F_{amp}). Then, the plates were inverted and incubated for 16 to 24 hours at 36°C±1°C. The concentrations of coliphage were calculated based on the number of circular zones of clearing (plaques) in lawns of host bacteria.

4.2.3 Results

Tables 4-1 and 4-2 show inactivation of male-specific coliphage and somatic coliphage by LP UV irradiation in the wastewater matrices. The inactivation of both somatic and male-specific coliphages in Matrix 3 samples was significantly lower than those in the Matrix 1 and 2 samples. The extent of coliphage inactivation in Matrix 3 samples for each UV dose was also significantly different from that in Matrices 1 and 2 (based on one-way ANOVA with 5% significance in SPSS 11.0 for Windows). The difference in the extent of inactivation between Matrix 3 and the other two matrices was sometimes more than 1 $\log_{10}$, especially at higher UV doses. These results indicate that aggregation and/or particle association shielded the associated coliphage and protected them from exposure to LP UV irradiation, thereby significantly reducing the UV disinfection efficacy. Therefore, UV disinfection systems in wastewater treatment processes should be optimized to compensate for possible reduction of disinfection efficacy due to extensive aggregation and/or particle association of viruses and perhaps other microbes.

Table 4-1. Inactivation of Male-specific Coliphages by LP UV Irradiation in Different Wastewater Matrices*.

Log$_{10}$ Reduction									
	Experiment 1			Experiment 2			Experiment 3		
UV Dose (mJ/cm^2)	Matrix 1	Matrix 2	Matrix 3	Matrix 1	Matrix 2	Matrix 3	Matrix 1	Matrix 2	Matrix 3
3	1.3	1.0	0.9	1.6	2.3	0.6	2.1	1.5	0.5
10	2.1	1.9	1.1	2.7	3.4	1.1	2.5	2.4	1.1
30	3.1	3.0	2.5	3.2	3.7	2.4	>3.5	3.2	1.4

*Matrix 1: supernatant, Matrix 2: settled, Matrix 3: resuspended

Table 4-2. Inactivation of Somatic Coliphages by LP UV Irradiation in Different Wastewater Matrices*.

Log$_{10}$ Reduction									
	Experiment 1			Experiment 2			Experiment 3		
UV dose (mJ/cm^2)	Matrix 1	Matrix 2	Matrix 3	Matrix 1	Matrix 2	Matrix 3	Matrix 1	Matrix 2	Matrix 3
3	0.9	0.2	-0.2	1.2	0.9	0.2	1.3	1.1	0.0
10	2.2	1.7	0.4	1.8	1.7	0.7	2.3	2.2	0.1
30	2.9	2.7	1.5	2.9	2.4	1.4	2.4	4.1	1.1

*Matrix 1: supernatant, Matrix 2: settled, Matrix 3: resuspended

4.3 UV Inactivation of Isogenic Pairs of Bacteria

4.3.1 Introduction and Purpose

Paired, isogenic strains of *S. typhimurium,* and *Acinetobacter* differing in key morphological or physical properties that may influence disinfection resistance were studied. *S. typhimurium* WG45 (female) and WG49 (F+) constitute one pair that differ in strain. WG49 contains an F-plasmid that confers antibiotic resistance (kanamycin), and the male (F+) trait that is responsible for the production of F-pili. These pili are surface appendages that may act as barriers against chlorine and UV disinfection. The *Acinetobacter* are an isogenic pair, differing by aggregated versus dispersed growth, and therefore, were expected to show differences in response to disinfection due to differences in this physical state. The response of the *Acinetobacter* strains to disinfection, specifically monochloramine disinfection, has been previously studied (Stewart, 1983, 1991).

4.3.2 Materials and Methods

4.3.2.1 Bacterial Strains

The strains of *Acinetobacter* EB22 and P7WT and *S. typhimurium* WG45 and WG49 were obtained from Pat Russin at the Unversity of Arizona, and Dr. James D. Oliver at the UNC–Charlotte, respectively. Seed stock organisms were received in agar slants, which were then used to inoculate growth media for experimental stocks. Purity of *Salmonella* cultures was verified by plating organisms on selective *Salmonella/Shigella* agar (Difco SS agar #0074-17). Experimental stocks were grown to mid-log phase in nutrient broth and preserved by adding a 20% glycerin solution (FisherChemicals #G33-500) before freezing down to –10°C for storage.

Acinetobacter strains were grown in tryptic soy broth (Difco #0370-17) and on tryptic soy agar (Difco #0369-17). *Salmonella* were grown in tryptic soy broth (TSB) without dextrose (Difco #0862-17) and on TSA (Difco #0368-17). For *Salmonella*, nalidixic acid (Sigma #N4382) was added to all media in 100x concentrations (1 mL per 100 mL media), and 0.2μm disc-sterilized kanamycin (Shelton Scientific #IB02120) was added to all *WG49* media in 100x concentrations. Tests were done to verify that nonresistant *Salmonella WG45* would not grow in media with kanamycin.

4.3.2.2 Quantification of Bacteria in Solution

Growth curves were performed on all bacteria prior to irradiation experiments to ensure consistent colony counts in exposure solutions. To perform growth curves, overnight inoculation cultures were started by adding two loopfuls of frozen stock from a cryotube to 50 mL of TSB in a capped, sterile 250-mL Erlenmeyer culture flask. These cultures were placed in a nonshaking water bath at 26°C for 24 hours (*Acinetobacter*) or 21 hours (*Salmonella*). Inoculation cultures were then diluted with sterile phosphate buffered saline (PBS, EM Science #6505) to an absorbance of approximately 0.10 optical density at 420 nm (Cary 100 Bio Spectrophotometer). This solution was used to inoculate test cultures. Test cultures consisted of 100 mL TSB into which 1.0 mL of the diluted inoculation culture was added before being placed in a shaking water bath at 37°C.

To determine the number of cfus per milliliter of broth media (cfu/mL), samples of the test broth solution were removed from the flask at specific time intervals, serially diluted, and then plated onto agar media. An absorbance reading with the spectrophotometer at 420 nm was recorded immediately before samples were diluted and plated so that correlations between time, cfu/mL and absorbance could be compared. Serial dilutions (10x) were made so that individual colonies could be counted on the Spot Plate assay. Serial dilutions in all experiments were conducted with sterile PBS as the diluting agent. Fifteen-milliliter centrifuge tubes (Corning) were used for each dilution, with each tube washed, rinsed and heat sterilized two times after opening before final disposal. Tubes were vortex mixed after the addition of each 1.0-mL dilutant, and then again before samples were spotted onto agar plates. PBS solution was added to each tube under a laminar flow hood with a sterilized repeating pipettor.

The spot plate technique permitted quantification of bacterial numbers in a nutrient broth solution. The spot plating agar dish was divided into four quadrants, onto which five 10 ul drops of each test dilution were placed per quadrant using a micropipettor. All work was done under a HEPA laminar flow hood, and spots were allowed to completely dry before the dishes were capped and placed upside-down in an incubator at 37°C.

Acinetobacter EB22 and P7WT plates were incubated for 10 to 12 hours at 37°C. *Salmonella* WG45 was incubated for 6 to 7 hours, with WG49 requiring 12 to 15+ hours before colonies grew enough to be effectively counted. In cases where more than one dilution had a countable number of colonies, the lower dilution (greater number of colonies per spot) was chosen. Colonies were tallied with the aid of a lighted Quebec colony counter magnified stage.

Colonies in each of the five spots of the selected dilution were then counted and input into a datasheet for conversion into cfu/milliliter. This datasheet averaged the number of colonies in all five 10 ul replicate spots, multiplied this number by 100 to convert to milliliters, and then multiplied again by the dilution factor to give cfu/milliliter.

4.3.2.3 UV Irradiation Setup

All irradiance experiments were conducted using an LP UV lamp inside of a bench-scale apparatus described earlier. UV irradiance from the lamp was measured using a radiometer (International Light, IL1700) calibrated at 254 nm. The radiometer sensor was placed at the center of the UV beam beneath the collimating tube at the height of the liquid surface of a 60 mm petri dish filled with 10 mL (approximately 0.509 cm) of water. Dose values for the lamp were calculated correcting for absorbance and non-homogeneity of light across the irradiation surface area. UV dose was calculated as the product of irradiance and time (Dose (mJ/cm^2) = I x t) where I = irradiance in mW/cm^2, t= time in seconds.

4.3.2.4 Experimental Protocol

All irradiance experiments were conducted on bacteria cultured according to accepted methods. Cultures were incubated in a shaking water bath at 37°C for the approximate time indicated by growth curve experiments. Samples were removed from the incubator at approximately 10^{10} cfu/mL, diluted two-fold (10^8) into refrigerated sterile PBS (5 mL broth to 495 mL PBS), and placed in an ice bath. Fluids were kept cold throughout experiments to limit further growth and maintain known cell numbers.

Exposures were conducted by placing 10 mL of the bacterial suspension in a 60-mm petri dish on a magnetic stirring platform under the collimating tube and exposing it for an appropriate amount of time to achieve a given UV dose. Two replicates were done for each dose value per bacterial strain in a randomly determined pattern. Dose values ranged from 0 to 40 mJ/cm^2, with 0 doses treated exactly like exposure doses, except with a closed shutter over the UV light source. A control 0 dose was also plated away from the lamp, along with controls to verify the sterility of TSB and PBS solutions at various points throughout the experiments.

After irradiation, samples were kept in the petri dishes, covered and held in the dark to limit possible photoreactivation before plating. If 24 total samples were exposed, plating was done after the first 12 were completed. Exposure solution contents were poured out of the petri dishes into a sterile 15-mL centrifuge tube, vortex mixed, serially diluted, and plated onto TSA with antibiotics or nalidixic acid if necessary using a spot plating protocol. Plates were then incubated overnight, colonies counted, and results tabulated.

4.3.3 Results

It was difficult to ascertain which *Acinetobacter* strain, P7WT or EB22, was the aggregating organism. A microscopic investigation was conducted of both strains grown under conditions expected to induce aggregation based on communications from the originator of the strain. The hypothesis was that the aggregating strain would be more resistant to UV light exposure. The relative degree of aggregation was higher in the P7WT, but it was difficult to ascertain the degree of aggregation due to low visible cell numbers in a given field of view, and little cell aggregation of significance was observed. These poor results in attempts to aggregate these cells minimized the level of experimentation that was performed.

Results of UV exposure experiments for *Acinetobacter* EB22 and P7WT, and *Salmonella* WG45 and WG49 generally showed now differences in the pairs of bacteria (data not shown). For *Acinetobacter,* eight UV doses from 0 to 40 mJ/cm^2 were selected for investigation. Three-log inactivation was reached in both strains at a dose of 15 mJ/cm^2. Although in one experiment EB22 showed greater resistance to UV exposure than P7WT, these results were not consistent across all experiments. For *Salmonella*, doses ranged from 0 to 20 mJ/cm^2. Three-log inactivation was reached in both strains above doses of 12 mJ/cm^2. At dose values <5 mJ/cm^2, both *Salmonella* strains show equal resistance to inactivation by UV light. At higher doses, non-antibiotic resistant WG45 showed a slight increased resistance as compared to kanamycin-resistant WG45, but this difference was not deemed significant.

Although some aggregation was observed among the *Acinetobacter* strains, the extent was not significant enough to consistently impact the UV disinfection effectiveness under the conditions tested. Furthermore, the difference in antibiotic resistance between the *Salmonella* strains tested did not appear to impact the effectiveness of UV disinfection. Thus, the hypothesis that the presence of the antibiotic-resistant plasmid increases resistance to UV irradiation was not supported. Finally, because bacteria are relatively susceptible to UV light and FC, establishing differences over small levels of disinfectant doses is challenging. However, the findings are important in that for these bacteria, there was no particular resistance conferred due to the morphological differences.

4.4 FISH for Identification of Pathogens Associated with Particles and Flocs in Wastewater

4.4.1 Introduction and Purpose

Although it is easier to demonstrate the concern of particle association using coliform as a model microbe, it is also necessary to establish if actual pathogens associate with particles in a similar fashion. However, because most pathogens are not present in numbers high enough to detect particle association using methods utilized for coliform above, a different approach was necessary to establish if pathogens were also associating with particles. The approach used was to implement some molecular biological tools to identify pathogens in wastewater and establish the degree of particle association using visualization methods. The method chosen for identification of pathogens associated with particles and flocs in wastewater was fluorescent in-situ hybridization (FISH).

Application of in-situ hybridization in environmental samples presents a unique set of problems. Environmental samples are likely to cause background fluorescence due to their organic and inorganic contents. By choosing a proper labeling technique or a fluorescent label, the background fluorescence of a sample can be reduced significantly, enabling the visualization of a bright signal from the microbe under investigation. Preliminary experiments have shown that background fluorescence is of particular concern in studying microorganisms associated with particles, since it is difficult to differentiate a positive microbe from naturally fluorescing particles.

In addition, methods chosen for fixation and hybridization are very important for in-situ hybridization of particle-associated microorganisms. Both fixation and hybridization involve steps that may affect the structure and size of wastewater flocs, which in turn may change the association of microorganisms with particles. Therefore, direct application of FISH techniques developed for pure cultures are not appropriate for the purpose of this type of research. In this study, a number of FISH methods developed for water, wastewater, sludge, and soil have been studied, and the appropriate techniques were modified and tested for the determination of particle-associated microbes. The main goal was to retain the original spatial distribution of the microbes without damaging the structure of the wastewater flocs and particles while conducting FISH.

The following FISH probes were used to study the particle association of all bacteria, *E. coli*, *S. typhimurium*, and *C. parvum*.

- EUB338 (16S rRNA) for all bacteria
- EC1531 (16S rRNA) for *E. Coli*
- Sal3 (23S rRNA) for *Salmonella*
- Cry1 (18S rRNA) for *C. parvum*

These are the most commonly used probes. The hybridization reactions are specific to the targeted microorganisms and the rRNA sequences.

Various fluorochromes and labeling techniques were tested for their suitability in this research:

- 5` tetramethyl rhodamine (TMR)
- 5` CY3
- 3` GreenStar*^{TM} –fluorescein isothiocyanate (FITC)

TMR is one of the most commonly used fluorochromes. Both probe-conferred signal from the targeted microbes and autofluorescing particles fluoresce in red when TMR is used. GreenStar*^{TM} –FITC (GeneDetect.com, New Zealand) labeled probe is prepared using a recently developed labeling technology which chemically attaches a cross-linked structure of multiple molecules of FITC to the 3' end of the probe. Use of multiple molecules of the fluorescent label greatly enhances the intensity of the positive signal and may make it easier to differentiate the targeted microbes from the autofluorescing particles. Another way to differentiate the microorganisms from the particles may be the use of a fluorescent dye, such as CY3, that produces much brighter, more photostable signal, and emits less background than most other fluorophores. When samples treated with a CY3 labeled probe are viewed with CY3 filter set, targeted microorganisms fluoresce in bright yellow, and debris fluoresces in red, which makes it possible to differentiate the particles from the microorganisms.

It is essential to use controls with in-situ hybridization experiments. Controls are needed to ensure that the hybridization reaction is specific, and that the probe specifically binds to the targeted rRNA sequence. In addition, if no response is present, it is important to know if this is because of an experimental error or the absence of the targeted microorganisms in the sample. The following controls were used during the experiments:

- All probes were tested with pure cultures.
- Pure cultures were also used at each run as a control.
- Poly(dT) probe was used for the confirmation of the mRNA in the samples.
- All samples were double stained with DAPI.
- Non-sense probes were used for nonspecific binding.

There are a number of different ways that in-situ hybridization experiments can be conducted. For the purpose of this research, it was essential to maintain the original spatial distribution and number of microorganisms associated with particles. This required modifications in the fixation, hybridization, and wash steps to minimize the disturbance to the floc structure and particle association.

The FISH techniques that have been used in this study can be classified in two main groups: (1) FISH on microscope slides; and (2) FISH on filters. Modifications were applied to each technique to minimize the possible changes to the particle association and improve image clarity. The procedures used are explained below.

4.4.2 Materials and Methods

4.4.2.1 FISH on Microscope Slides

The method developed by Amman et al. (1995) was used to hybridize fixed wastewater sample on microscope slides. Four-percent paraformaldehyde solution was used to fix the freshly collected wastewater sample by adding three volumes of fixative to one volume of sample. The mixture was kept at 4°C for 3 hours, followed by centrifugation at 5000xg to remove the fixative. Cells were washed, suspended, and centrifuged with 1xPBS twice. One volume of ice-cold 70% ethanol was added to complete the fixation. Then, 10 μL of the fixed cells were spread on each well of the gelatin coated slides, and the smears were allowed to air-dry overnight. The next morning, the smears were dehydrated by immersing slides into a series of 50%, 80%, and 96% ethanol for 3 minutes each.

Hybridization buffer contained 0.9 M NaCl, 20 mM Tris-HCl, 0.01% sodium dodecylsulphate (SDS), and 20% to 35% formamide, depending on the probe used. For each smear to be hybridized, 9 μL hybridization buffer was mixed with 1 μL probe (25 ng/μL) on parafilm, and added to the wells. The slides were carefully placed in prewarmed polypropylene centrifuge tubes, and hybridized at 46°C for 3 hours. After hybridization, the slides were rinsed with wash buffer that contained 20 mM Tris-HCl (pH=7.2), 10 mM EDTA, 0.01% SDS, and 308-80 mM NaCl, depending on the formamide percent used in the hybridization buffer. Then each slide was placed in centrifuge tubes with prewarmed 50-mL hybridization wash buffer, and incubated for 20 minutes at 48°C. Slides were quickly dipped in DI to remove the salts and air-dried. Then, 5 μL of DAPI (1 mg/mL) was applied on each well for 5 minutes and washed off with DI. After the slides were air-dried in the dark, 1 μL of antifade solution was applied to each well, and the slides were covered with cover glass. The slides were sealed with clear nail polish, and visualized under the microscope.

Preliminary experiments have shown that the use of the FISH method to detect particle-associated microorganisms may not be appropriate. The fixation step involves repeated centrifugation, which is very likely to change the floc structure, particle size, and particle-association of microbes. Therefore, modifications to the fixation step were needed. Instead of removing the fixative using centrifugation, the fixed sample was filtered through an 11-μm nylon filter, and the particles were gently resuspended in 1xPBS. The majority of the particle-associated microbes are associated with particles larger than 10 μm (Emerick et al., 1999). Therefore, loss of particles <11 μm due to the filtration process should not change the results significantly. The nylon filter works very well for this purpose since it does not absorb interstitial water from the flocs and does not stick to particles. Although there can be some disruption due to resuspension, particles can be suspended in 1xPBS after filtration without causing extensive changes in their structure or microbial association compared to centrifugation. Polycarbonate and cellulose filters were also used with smaller pore sizes for filtration. However, particles adhere firmly to these filters and cannot be resuspended easily.

4.4.2.2 FISH on Membrane Filters

In order to preserve the original floc structure and particle association of microbes in wastewater, it might be better to do FISH directly on membrane filters rather than on microscope slides. Two types of membrane filters are commonly used in the field for this purpose: (1) black polycarbonate filter; and (2) white polycarbonate filter. Polycarbonate membrane filters have a very smooth surface that makes them ideal for FISH experiments. Due to firm adhesion to polycarbonate filters, only about 10% of cells are lost during hybridization and washing (Glockner et al., 1996). The nylon filters as used by Emerick et al. (1999) were also tested.

Various combinations of fixation and hybridization techniques were tested with membrane and nylon filters to determine a FISH protocol that would cause minimum damage to the floc structure and particle association and yield clear images that would allow quantification of microorganisms and observation of their spatial distribution. These techniques are listed below.

Hybridization was conducted on the filter. The fixed wastewater sample (fixation procedure is same as explained above) was filtered through the filter. PBS and DI were then filtered to wash off the excess fixative. The filter was placed on a microscope slide, hybridization buffer and probe were laid on the filter, and the filter was hybridized for 3 hours. Then, the filter was placed in prewarmed wash buffer for 20 minutes, and rinsed with DI water. DAPI and antifade solutions were applied before visualization.

Fixation and hybridization were directly conducted on the filter. Wastewater sample was filtered, and the filter was overlaid with 4% fixative for 30 minutes. The fixative was removed by vacuuming, and PBS and DI water were vacuumed subsequently to remove the excess fixative. The filter was placed on a glass slide, and hybridization was carried out as explained above. After hybridization, the filter was washed in the wash buffer, and DAPI and antifade solutions were applied before visualization.

Fixation and hybridization were directly conducted on the filter when the filter was still in the filtration unit. Only before visualization, the filter was removed from the filtration unit and placed on a microscope slide. This technique minimized the handling of the filter.

The sample was fixed and hybridized in solution, which was concentrated on the filter. The filter was washed in the wash buffer, rinsed with DI, and placed on a glass slide. DAPI and antifade solutions were applied before visualization.

4.4.3 Results

The results of this study indicate that not only bacteria, but also *C. parvum* oocysts, are associated with particles. This study specifically illustrated the particle association of bacteria, *E. coli*, *Salmonella*, and *C. parvum*. These results are displayed in Figures 4-2 through 4-5.

Two important challenges are present with in-situ determination of particle-associated microorganisms: (1) to maintain the original floc structure and particle association of microbes during the FISH protocol; and (2) the background fluorescence caused by wastewater constituents that can be mistaken with probe signal from the targeted microbes. However, both of these difficulties can be minimized by choosing the right label and filter set and by modifying various techniques used during FISH.

Differentiation of autofluoresceing particles from microbes is easier if probe-conferred signal is in a different color than the particles. Use of CY3 with the CY3 filter set enables visualization of the microbes in orange-yellow and particles in red. If both particles and microbes fluoresce in same color (as with TMR and FITC), it is almost impossible to tell what is a microbe and what is a particle. Use of multiple molecules of the fluorescent label enhances the intensity of the positive signal, but it also seems to enhance the background fluorescence.

It is not recommended to use centrifugation during the fixation step. Centrifugation changes the structure and size of particles and flocs, and thus changes the particle association of microbes. Fixative can be removed successfully without damaging the particles by employing a nylon filter.

FISH on microscope slides and membrane filters works well for visualization of particle-associated microorganisms as long as care is taken to preserve the floc structure and related particle association of microbes.

White polycarbonate filters seemed to work much better than black polycarbonate filters. Black polycarbonate filter bleaches during the FISH protocol, and patches of blacks and grays are observed under the microscope. Use of nylon filter is not recommended since the sample cannot be uniformly distributed over the filter and will be collected between the nylon fibers.

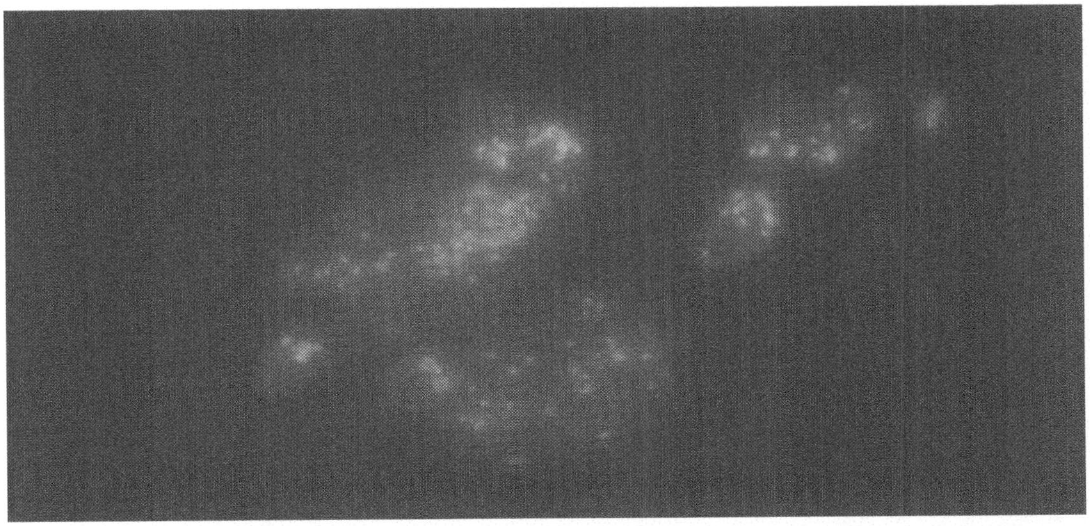

Figure 4-2. Image of Wastewater Bacteria Associated with Particles and Flocs, Unedited. The probe used is EUB338 labeled with CY3. The image is viewed using a CY3 filter set. The bacteria fluoresce in orange-yellow, and background fluorescence in red. Hybridization was conducted on a glass microscope slide. Magnification is x1000.

Figure 4-3. Image of Indigenous Wastewater *E. coli* Associated with Particles and Flocs, Unedited. The probe used is GreenStar–FITC labeled EC1531. Use of multiple FITC molecules increases the intensity of the positive signal from *E.coli*. However, background fluorescence is still a problem. Note the large particle in the middle of the floc that has a signal as bright as the *E.coli*. Hybridization was conducted on a glass microscope slide. Magnification is x1000.

Figure 4-4. Image of Wastewater Sample Hybridized on a Nylon Mesh Filter (11 µm pore size) and Stained with DAPI, Unedited. As seen from the image, the filtered sample accumulates between the nylon fibers of the filter, and is not distributed evenly over the filter surface. Magnification is x400.

Figure 4-5. Image of *C. parvum* oocysts Associated with Wastewater Particles, Unedited. The sample was hybridized on a white polycarbonate membrane filter. The probe is CRY1 that is labeled with CY3. Magnification is x1000.

CHAPTER 5.0

OPTIMIZING APPLICATION OF DISINFECTANT DOSES FOR PATHOGENS IN WASTEWATER

5.1 Introduction

In complex matrices such as wastewater, many factors can impact the disinfection of microorganisms. It is important to understand how the performance of certain disinfectants is impacted by the water matrix and what strategies can be implemented in the procedures for disinfection application to optimize use of the disinfectant for achieving the water quality goals. In this phase of the research, variations in the application of chlorine and UV disinfectants were investigated, and scenarios were evaluated for optimizing the effectiveness of wastewater disinfection.

5.2. Alternation of Concentration and Time Within the Same Chlorine CT Values

5.2.1 Introduction and Purpose

The overall goal of this research phase was to optimize wastewater disinfection processes in order to better protect the public from exposure to harmful effects of various waterborne pathogens. Although wastewater disinfection has been practiced for decades, there has been very little effort to optimize the disinfection systems in terms of effective inactivation of different types of waterborne microorganisms. This is probably because wastewater treatment plants (WWTPs) in the United States are required by federal and state law to demonstrate their ability to reduce the level of only one type or group of microorganisms—coliforms or fecal coliforms—under a certain level (e.g., 100 coliforms/mL). Current chlorination practice is effective against coliforms for a wide range of concentrations (5 mg/L to 30 mg/L) and contact times (a few minutes to hours) in wastewater treatment processes. However, little is know about the relative effectiveness of these different chlorination practices in wastewater against other types of microorganisms, such as enteric viruses, bacterial spores, and protozoan parasites. In general, viruses, bacterial spores, and protozoan cysts and oocysts are more resistant to chlorine and other disinfectants than fecal indicator bacteria, such as fecal coliforms and *E. coli*. Therefore, the effect of variations in concentration and exposure time at equal chlorine CT (dose or concentration * exposure time) values on inactivation of *C. perfringens* spores, somatic and male-specific coliphages, and *C. parvum* oocysts was determined.

5.2.2 Materials and Methods

5.2.2.1 Wastewater

Wastewater samples were taken from the Morehead City sewage treatment plant (Morehead City, North Carolina). This facility has primary physical and secondary biological process (trickling filtration). The secondary effluent is disinfected using a chlorination process (liquified chlorine gas). Samples of wastewater for experiments were taken before the chlorination unit. The concentration of ammonia in these wastewaters was approximately 3 mg/L with a pH ranging from 7.6 to 7.8.

5.2.2.2 FC Solutions and Measurement of Chlorine Residuals

Household bleach (6% sodium hypochlorite; Clorox) was diluted in distilled water to prepare a 1000 mg/L stock solution. Total chlorine and FC residuals were measured by Hach Chlorine Test Kits (Model AccuVac Kit (for FC) and CN-21P (for total chlorine)) which are based on the DPD colorimetric method. The working ranges of these kits are 0 to 2.5 mg/L and 10 mg/L to 100 mg/L. The smallest increments are 0.1 mg/L and 10 mg/L for AccuVac Kit and CN-21P kit, respectively. These Hach kits have been used in many wastewater treatment plants, and the AccuVac Kit has been accepted by the EPA for reporting purposes.

5.2.2.3 Experimental Protocol for Chemical Disinfection with FC

Chlorine disinfection experiments were done in a bench-scale, batch system using three polypropylene bottles that can hold ~250 mL of wastewater. After taking a sample from the negative control bottle for determining the initial phage and *C. perfringens* spore levels in the wastewater, predetermined amounts of FC stock solution were added to the other two bottles to achieve two different initial FC concentrations (10 mg/L and 20 mg/L). Then, samples were allowed to react for a certain contact time to give two target CT (concentration * time) values of 100 and 1000 mg-min/L (10 minutes and 100 minutes for 10 mg/L and 5 minutes and 50 minutes for 20 mg/L). Samples were withdrawn from the reaction vessels at time=0 and at the four target times. Each sample was divided into two portions. One sample was used directly for chlorine residual measurement; the other was immediately dechlorinated with 0.1% sterile sodium thiosulfate and assayed for coliphage by SAL method and for *C. perfringens* by membrane filtration method.

5.2.2.4 Microbial Assays

Somatic and male-specific coliphage were assayed as in an EPA method (Method 1602: Male-specific (F+) and somatic coliphage in water by SAL procedure). Briefly, both test samples of wastewater and bacterial hosts grown overnight (CN13 for somatic and F_{amp} for male-specific coliphage) were added to 2X TSA-containing antibiotics (nalidixic acid for CN13 and ampicillin/streptomycin for F_{amp}). Then, the plates were inverted and incubated for 16 to 24 hours at 36°C±1°C. The concentrations of coliphage were calculated based on the number of circular zones of clearing (plaques) in lawn of host bacteria. *C. perfringens* spores were assayed by membrane filtration method on mCP agar plates. Both control and disinfected samples were filtered through a membrane and the membrane was mounted on mCP agar plates. Then, the plates were incubated in an anaerobic chamber at 41°C for 18 to 24 hours. *C. parvum* infectivity assay was done as described in the previous section.

5.2.3 Results

Tables 5-1 and 5-2 show the log inactivation of *C. perfringens* spores and coliphages by 10 mg/L and 20 mg/L of initial FC in the wastewater. The wastewater effluent used for these experiments contained large amounts of ammonia (i.e., this was a non-nitrified effluent and the FC added immediately reacted with the ammonia [and possibly other constituents of the wastewater] to form chloramines [the residual FC level dropped quickly, ~1 mg/L within 1 minute and usually ~0.5 mg/L at the end of contact time {10 minutes}]). In essence, this was not actually a chlorination process, but a chloramination process. That is, added FC reacted with ammonia and much of the added FC was relatively quickly converted to chloramines. Nonetheless, this approach was utilized because this is a common method of application (dosing FC into wastewater) practiced by many WWTPs. One of the project's goals was to determine the efficacy of current disinfection practice on inactivation of some important indicator microorganisms.

The inactivation of *C. perfringens* spores by FC was relatively slow and only ~0.3 log inactivation was achieved with a total chlorine CT value of 100 mg/L*min. However, the inactivation level increased to more than 1 log with a CT value of 1000 mg/L*min. The inactivation of somatic coliphages by FC was relatively fast and at least ~1 log and ~2 log inactivation was achieved with total chlorine CT values of 100 mg/L*min and 1000 mg/L*min, respectively. However, the inactivation of male-specific coliphages was relatively slow and only ~ 0.5 log and ~1 log inactivation was achieved with total chlorine CT values of 100 mg/L*min and 1000 mg/L*min, respectively (except in Experiment 1). Although the levels of microbial inactivation by dosing FC varied a lot in each experiment (due to differences in the quality of wastewater) and the number of experiments were smaller than ideal (due to time and personnel constraints), the inactivation of each microorganism by the same total chlorine CT values (100 mg/L*min and 1000 mg/L*min) were not significantly different at the 5% level by paired two sample t-tests. Although in theory identical CT values would result in similar inactivation of *C. perfringens* spores and coliphages regardless of the different levels of chlorine concentration and contact time, a larger data set would be desirable to draw definitive conclusions. However, this dataset does point out some important trends and provides insights into these issues. Table 5-3 shows the log inactivation of *C. parvum* oocysts by the same concentrations of initial FC in wastewater. Although there was very low level of inactivation in FC disinfection conditions, it appears that identical CT values from different initial chlorine concentrations would result in similar inactivation of *C. parvum* oocysts as well.

Table 5-1 Inactivation of Microorganisms by FC (CT=100 mg/L*min) in Wastewater.

		Log_{10} Reduction								
Conditions		*C. perfringens*			Somatic coliphages			Male-specific coliphages		
mg/L	min	Exp. 1	Exp. 2	Exp. 3	Exp. 1	Exp. 2	Exp. 3	Exp. 1	Exp. 2	Exp. 3
10	10	0.36	0.29	-0.12	2.01	1.35	1.48	3.07	0.99	0.70
20	5	0.28	0.33	-0.09	2.13	0.88	0.75	0.54	0.48	0.34

Table 5-2 Inactivation of Microorganisms by FC (CT=1000 mg/L*min) in Wastewater.

Log10 Reduction										
Conditions		*C. perfringens*			Somatic coliphages			Male-specific coliphages		
mg/L	min	Exp. 1	Exp. 2	Exp. 3	Exp. 1	Exp. 2	Exp. 3	Exp. 1	Exp. 2	Exp. 3
10	100	0.81	2.22	0.76	3.11	1.84	1.91	3.15	1.07	0.89
20	50	2.21	1.84	0.97	3.19	1.56	1.26	0.78	0.80	0.47

Table 5-3 Inactivation of *C. parvum* Oocysts by FC (CT= 100 and 1000 mg/L*min) in Wastewater.

Conditions		Log_{10} Reduction in *C. parvum* oocysts
CT=100		
10 mg/L	10 min	0.05
20 mg/L	5 min	0.02
CT=1000		
10 mg/L	100 min	0.03
20 mg/L	50 min	0.13

5.3 Reciprocity of UV Irradiance and Exposure Time for Inactivation of *E. coli*

5.3.1 Introduction and Purpose

The same product of irradiance and time (time-dose reciprocity) may not lead to the same reduction of microorganisms when different levels of irradiance are used. UV inactivation does not always follow the time-dose reciprocity for water disinfection. Sommer et al., (1996) reported that in several eukaryotic yeast strains, low intensity over long exposure times are more effective in inactivating cells than higher intensity and shorter exposure times. In contrast, *E. coli* showed higher UV inactivation when applying a high UV irradiance over short exposure times (Sommer et al., 1998). Wiedenmann et al., (1997) concluded that in chemical disinfection (chlorine, ozone) the same concentration-time values might not lead to the same reduction of microorganisms.

5.3.2 Materials and Methods

The test organism used for this study was *E. coli* strain C3000. The *E. coli* was cultured in tryptic soy broth at 37°C. Microbial analysis of the test water was performed using a membrane filter technique, where the bacteria produced a red colony with metallic sheen within 24 hours of incubation at 35°C on an Endo-type medium (American Public Health Association, Standard Methods, 1995, 19th Ed, Method 9222 B).

5.3.2.1 Water Sample

A laboratory water developed in the lab to mimic a natural surface water was prepared. The water contained all of the major ions, had a pH of 6.8, and had a total organic carbon (TOC) of approximately 7 mg/L consisting of spiked natural organic matter and alginic acid.

5.3.2.2 LP UV Irradiation and Radiometry

A bench scale quasi-parallel beam UV apparatus consisted of LP UV mercury vapor germicidal lamps producing monochromatic UV radiation at 253.7 nm. UV irradiance at

253.7 nm was measured with an International Light radiometer and UV detector (Model IL1400A) with a sensor (Model SEL240).

5.3.2.3 Dose Determination

The incident irradiation at the surface was corrected to provide the true average incident irradiation across the top of the dish. The average irradiation in the solution was computed by the following parameters: the sample depth, the UV absorbance of the solution at 254 nm, and the true incident irradiation. The exposure time was then calculated by dividing the target UV dose by the average irradiation.

5.3.2.4 Irradiation Procedure

The *E. coli* was diluted into the water to a concentration of ~10^6/mL of solution. Then 10 mL of solution was placed in a 60-mm diameter petri dish, and irradiated with LP UV lamp, while stirring samples on a magnetic stir plate for predetermined exposure times. The petri dish was placed at the same distance from the UV lamp source through the whole experiment. Control samples were placed in a stirred petri dish for the same time as the exposed samples, without irradiation. After completion of all irradiated sets, the samples were serially diluted 10-fold to yield a countable number of *E. coli*. Each dose point was performed in duplicate.

5.3.2.5 UV Disinfection Experiment for Dose-Reciprocity Approach

An optical wire mesh (courtesy of Alexander Cabaj, University of Vienna) was fixed below the UV lamp, on top of the circular opening. The optical wire mesh enabled lower intensity (10-fold) at the surface of the sample. For this experiment each dose was irradiated in triplicate.

5.3.2.6 Data Presentation

Concentration (cfu/milliliter) of *E. coli* in the control sample was taken as the initial concentration N_0. For each dose the average of 3 replicates of *E coli* concentration was calculated, and was taken as N_d. The $\log_{10}$ transformation for N_d/N_0 (i.e. $\log_{10}(N_d/N_o)$) was plotted as a function of the dose.

5.3.3 Results and Discussion

The log inactivation of *E coli* resulting from varying the exposure time of LP UV lamp at irradiation of 0.286 mJ/cm^2 as a function of dose resulted in a 3 log reduction of *E coli* at dose of 10 mJ/cm^2, similar to UV inactivation data presented in the literature. As the detection limit of the assay was reached at approximately 5 log, a reduced rate of inactivation of *E. coli* was observed with increased dose of UV applied above 15 mJ/cm^2.

Table 5-4 presents the data set for the experiment examining log inactivation of *E. coli* for the same UV dose delivered using two different UV irradiance levels. In this experiment, the time-dose reciprocity was found to not hold true at a given UV dose. When *E. coli* was exposed to a dose of 10 mJ/cm^2 at high intensity (0.242 mJ/cm^{2}sec) the log inactivation was 2.83±0.12 however for a dose of 10 mJ/cm^2 using low intensity (0.024 mJ/cm^{2}sec) the log inactivation was 1.57±0.06. Thus a difference of ~1 log reduction was obtained at a dose of 10 mJ/cm^2 when applying high intensity of 0.242 mJ/cm^2 for a short time (without mesh) than applying a lower intensity of 0.024mJ/cm^2 over a long time (with mesh). Less then ±0.5 log variation resulted at each dose between replicates.

Sommer et al. (1998) examined three irradiation levels of 0.2 mW/cm^2, 0.02 mW/cm^2, and 0.002 mW/cm^2 and found similarly that for *E. coli* strains (ATCC 25922, ATCC 11229 and wild type) the time dose reciprocity was not valid. For each dose, the differences in the log reduction of the *E. coli* strains, between UV irradiance of 0.2 mW/cm^2 and 0.002 mW/cm^2, for doses of 8 mW-sec/cm^2, 9 mW-sec/cm^2, and 10 mW-sec/cm^2 were –1.23, -0.88, and –1.17, respectively. In contrast to these results with *E coli* strains, the other test organisms they observed (*B subtilis* spores, phage MS2, phage Φ174, and phage B40-8) did not differ from each other at the same dose levels, regardless of the level of UV irradiance applied. The present experiment provided comparable results to the study by Sommer et al. (1998) showing 1 log reduction difference for *E. coli* at dose of 10 mJ/cm^2 when applying 10-fold lower irradiance.

Table 5-4. Log Reduction (N_d/N_0) of *E. coli* at Two Different Irradiance Levels at a Dose of 10mJ/cm^2.

	Irradiance mW/cm^2	UV dose mJ/cm^2	Cell count 1	Log N_d/N_0	Cell count 2	Log N_d/N_0	Cell count 3	Log N_d/N_0	SD**
	0	0	2.00E+05	0	2.00E+05	0	2.50E+05	0	2.90E+04
No mesh	0.242	10	2.50E+02	-2.9	2.90E+02	-2.9	4.00E+02	-2.7	7.80E+01
With mesh	0.024	10	6.80E+03	-1.5	6.20E+03	-1.6	5.60E+03	-1.6	6.00E+02

* Average 0 dose is 2.20E+05
**SD = standard deviation

Sommer et al. (1998) explained that this effect was due to repair enzymes of the cell that were more influenced by high UV irradiance, which is opposite to the inactivation of eukaryotic yeast cells that are more susceptible to the time of exposure than to the UV intensity. In another study Sommer et al. (1996) explained these irradiance effects by different repair pathways, such as nucleotide excision repair, error-prone repair, and DNA double-strand repair compete for induced DNA damage upon different intensities. A further hypothesis would be the occurrence of the UV-sensitive components or regulators of the repair machinery.

This phenomenon may occur due to repair mechanism and is not due to regrowth at the lower intensities, over the exposure time. The time difference between low and high irradiance exposures (Table 4-4) was about 6 minutes, which is lower than the optimal doubling time of *E. coli*. One way this hypothesis can be tested is by performing an assay that quantifies the number of UV-induced pyrimidine dimers in the DNA such as that performed by Oguma et al. (2001). If regrowth of *E. coli* occurs, and not repair, then it would be expected that the number of pyrimidine dimers at high irradiance would be the same as low irradiance for the same time-dose reciprocity. If repair at low intensities were the issue, then it would be expected that the number of pyrimidine dimers at low irradiance would be less than the number of pyrimidine dimers at high irradiance, for the same time-dose reciprocity. Adoption of this assay for gaining insights into the UV inactivation mechanisms such as that found with time-dose reciprocity is a subject of future research.

5.4 Sequential Disinfection with UV and Chlorine

5.4.1 Introduction and Purpose

One goal of this research was to optimize wastewater disinfection processes in order to better protect the public from exposure to waterborne pathogens. According to previous research, a wide spectrum of microbial response against specific disinfectants has been observed, and responses differ from disinfectant to disinfectant. For example, although UV irradiation is very effective against protozoan parasites such as *Cryptosporidium* and *Giardia*, some enteric viruses are very resistant to UV disinfection. Chlorination is very effective against most enteric bacteria and some enteric viruses. However, *Cryptosporidium* and *Giardia* are extremely resistant to chlorination.

In order to control various types of waterborne pathogens, a prudent approach might be to apply various disinfectants in combination. Therefore, the effectiveness of several sequential disinfection on inactivation of the three major types of microorganisms, *C. perfringens* spores (bacteria), somatic coliphages (viruses), and *C. parvum* oocysts (protozoan parasites), was investigated. Specifically, this research tried to simulate a sequential disinfection scheme with chlorination ("in-situ" chloramination) followed by UV disinfection and another sequential disinfection scheme with (pre-formed) chloramination after UV disinfection. The same initial chlorine concentration (10 mg/L) was used in both sequential disinfection schemes, but different contact times of 10 minutes (for in-situ chloramination followed by UV) vs. 100 minutes (for UV followed by pre-formed chloramination) were used to better simulate each disinfection scheme. The results of this work may enhance the capability of a WWTP to reduce waterborne pathogens through use of multiple disinfectants and thereby better protect the public from exposure to these microbes.

5.4.2 Materials and Methods

The FC, measurement of chlorine residuals, experimental protocol for chemical and UV disinfection experiments, and LP UV dosimetry are presented above.

5.4.2.1 Wastewater

Wastewater samples were taken from sewage treatment plants in Morehead City, North Carolina, and Beaufort, North Carolina. These facilities have primary physical and secondary biological process (trickling filtration in Morehead City and activated sludge in Beaufort). The secondary effluents are disinfected with chlorination (liquified chlorine gas in both places). Samples for experimental disinfection in laboratory studies were taken before the chlorination units. The concentrations of ammonia were typically ~3 mg/L from the Morehead City plant and ~0.5 mg/L from the Beaufort plant.

5.4.2.2 FC and Measurement of Chlorine Residuals

Household bleach (6% sodium hypochlorite; Clorox) was diluted in distilled water to prepare a 1000 mg/L stock solution. Total and FC residuals were measured by Hach Chlorine Test Kits (Model AccuVac Kit and CN-21P) which are based on the DPD colorimetric method previously described.

5.4.2.3 Monochloramine

Monochloramine stock was prepared from FC and ammonia solution—1 mg of chlorine as HOCl and 4 mg of ammonia on a weight:weight ratio—by slowly mixing for up to 1 hour to generate a 100 mg/L stock. The yield was analyzed using the DPD colorimetric method.

5.4.2.4 Experimental Protocol for Chemical Disinfection Experiments

Chemical disinfection experiments were done in a bench-scale, batch system using polypropylene bottles which can hold ~250 mL of wastewater.

FC Followed by UV Irradiation: After taking a sample for determining the initial phage level in the wastewater, predetermined amounts of FC stock solution were added to the bottles to achieve an initial FC dose of 10 mg/L. Then, samples were allowed to react for 10 minutes and were divided into two portions. One portion was used directly for UV irradiation exposures and the other was immediately dechlorinated with 0.1 % sterile sodium thiosulfate. Coliphages were assayed using the SAL method and *C. perfringens* using the membrane filtration method. After UV irradiation, samples were also dechlorinated with sodium thiosulfate and assayed for test microorganisms.

UV Irradiation Followed by Monochloramine Disinfection: After taking a sample to determine the initial phage level in the wastewater, samples were irradiated with target UV doses. Then, samples were divided into two portions. One portion was used for following monochloramine disinfection and the other was directly assayed for coliphages by the SAL method and for *C. perfringens* by the membrane filtration method. For monochloramine disinfection, predetermined monochloramine stock solution was added to the bottles to achieve initial monochloramine dose of 10 mg/L and allowed to react for 100 minutes. A sample was withdrawn at the end of the contact time and was immediately dechlorinated with 0.1 % sterile sodium thiosulfate and assayed for test microorganisms.

5.4.3 Results

Figure 5-1 and 5-2 show the inactivation of indigenous *C. perfringens* spores, somatic coliphages, and spiked *C. parvum* oocysts by LP UV irradiation with or without prior chlorination at 10 mg/L for 10 minutes in two wastewater samples from Morehead City and Beaufort, respectively. Again, the wastewater effluent used for these experiments contained large amount of ammonia; i.e., this was a non-nitrified effluent, so that the FC added immediately reacted with the ammonia to form chloramines (more than 90% of the FC almost immediately reacted with the materials in the wastewater [including ammonia] and converted to chloramines). Due to time and personnel constraints, only one experiment was performed for each wastewater sample.

Inactivation of somatic coliphages by LP UV was very rapid and reached >3 $\log_{10}$ with a UV dose of 16 mJ/cm^2 with or without prior chlorine exposure (~1 $\log_{10}$ inactivation with 10 mg/L of FC for 10 minutes). Inactivation of *C. parvum* oocysts was also very rapid and reached ~2 $\log_{10}$ with a UV dose of 2 mJ/cm^2, although there was no apparent inactivation from prior chlorine exposure. Very little reduction of *C. perfringens* spores by LP UV and/or FC occurred, and <1 $\log_{10}$ inactivation was achieved with a UV dose of 40 mJ/cm^2 and prior chlorination (10 mg/L of FC for 10 minutes).

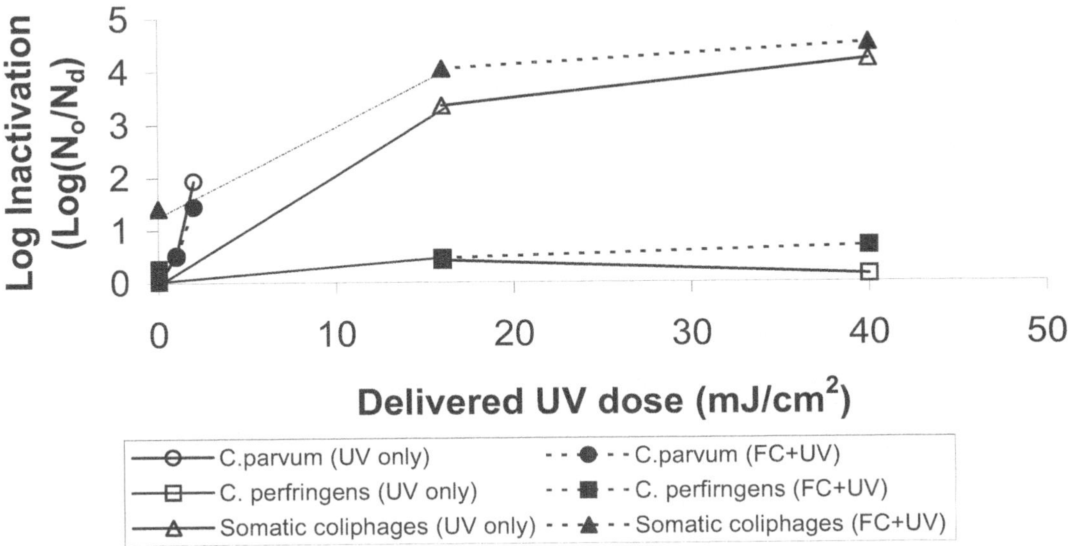

Figure 5-1. Inactivation of Microorganisms by Dosed FC (FC: 10 mg/L [initial concentration] for 10 minutes) and/or LP UV in Wastewater (Morehead City). [*The residual concentration of total chlorine, FC, and ammonia at each time point: At time = ~0 min, 10 mg/L, 0.4 mg/L, and >3 mg/L; at time = 10 min, 10 mg/L, 0.3 mg/L, 1.9 mg/L, respectively)].

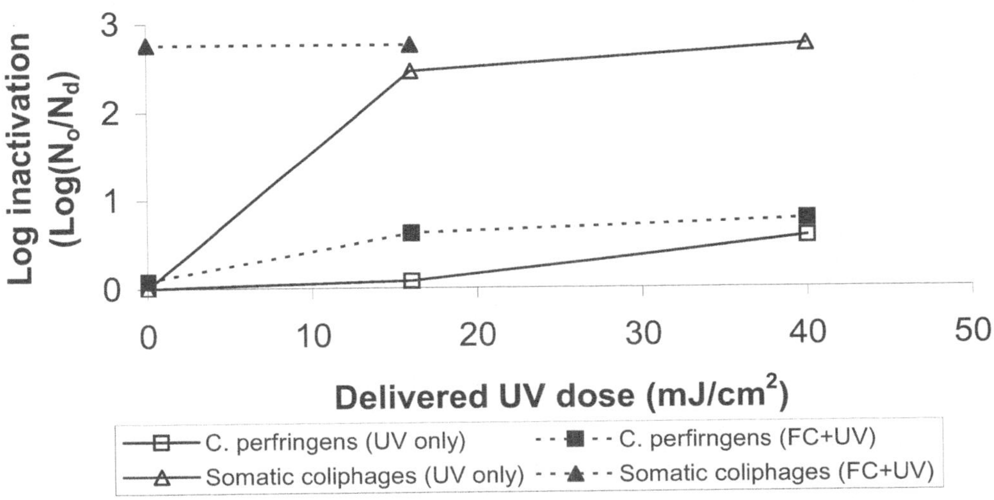

Figure 5-2. Inactivation of Microorganisms by Dosed FC (FC: 10 mg/L [initial concentration] for 10 minutes) and/or LP UV in Wastewater (Beaufort). [*The residual concentration of total chlorine, FC, and ammonia at each time point: At time = ~0 min, 10 mg/L, 1.7 mg/L; and 0.7 mg/L, at time = 10 min, 10 mg/L, 0.7 mg/L, 0.2 mg/L, respectively)].

Figures 5-3 and 5-4 show the inactivation of indigenous *C. perfringens* spores, somatic coliphages, and spiked *C. parvum* oocysts by 10 mg/L monochloramine with or without prior LP UV irradiation in two wastewater samples from Morehead City and Beaufort, respectively. Lower UV doses, 16 mJ/cm^2 for *C. perfringens* spores and somatic coliphages and 1 mJ/cm^2 for *C. parvum* oocysts, were used to maintain the levels of inactivation under the detection limits of the assays so that the presence and extent of synergism between UV irradiation and chloramination could be determined. Unlike sequential disinfection with FC and LP UV, inactivation of *C. perfringens* spores by monochloramine was rapid and reached ~3 $\log_{10}$ with 10 mg/L of monochloramine and 100-minute contact time even without prior LP UV exposure

(16 mJ/cm^2) that resulted in ~1 $\log_{10}$ inactivation. Inactivation of somatic coliphage was also relatively rapid and reached ~2 $\log_{10}$ with 10 minutes contact time in addition to 3 $\log_{10}$ inactivation by prior LP UV exposure (16 mJ/cm^2). (The initial titers of somatic coliphages in these wastewater samples were relatively low; only ~1/2 $\log_{10}$ additional inactivation was observed in 100 minutes of contact time.) However, there was very little reduction of *C. parvum* oocysts by monochloramine with 100-minute contact time with or without prior LP UV irradiation (1 mJ/cm^2). As mentioned earlier, a different UV dose (1 mJ/cm^2) was used for inactivation of *C. parvum* oocysts in this disinfection scheme (compared to the 2 mJ/cm^2 for in-situ chloramination and UV) in order to examine presence of synergism between these disinfectants for inactivation of *C. parvum* oocysts.

Because the detection limits of the infectivity assay for *C. parvum* was almost reached at UV dose of 2 mJ/cm^2 in earlier experiments (in-situ chloramination and UV), it would be difficult to examine the possible role of synergism between disinfectants under the former conditions. Therefore, a lower UV dose (1 mJ/cm^2) was used in later experiments (UV and pre-formed chloramination).

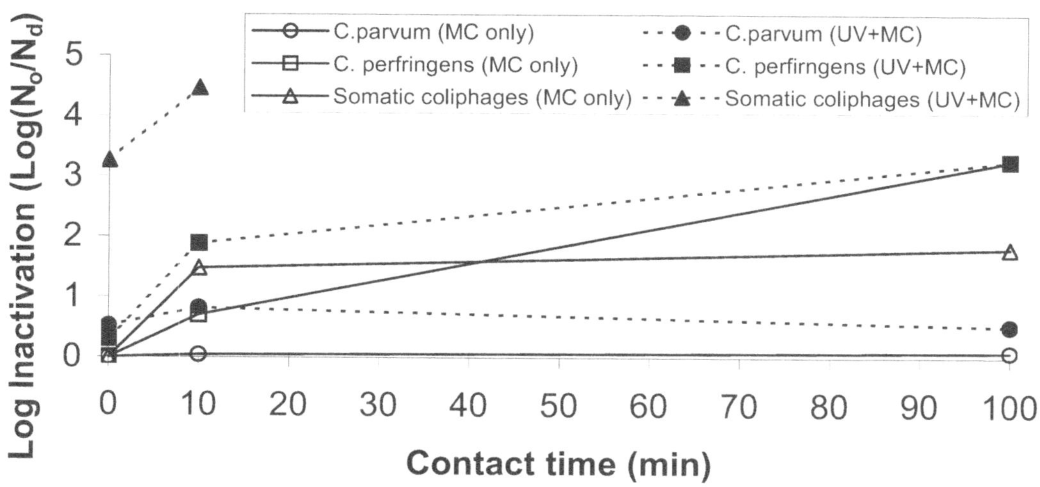

Figure 5-3. Inactivation of Microorganisms by UV and/or Monochloramine (Monochloramine: 10 mg/L [initial concentration] for 10 minutes) in Wastewater (Morehead City).

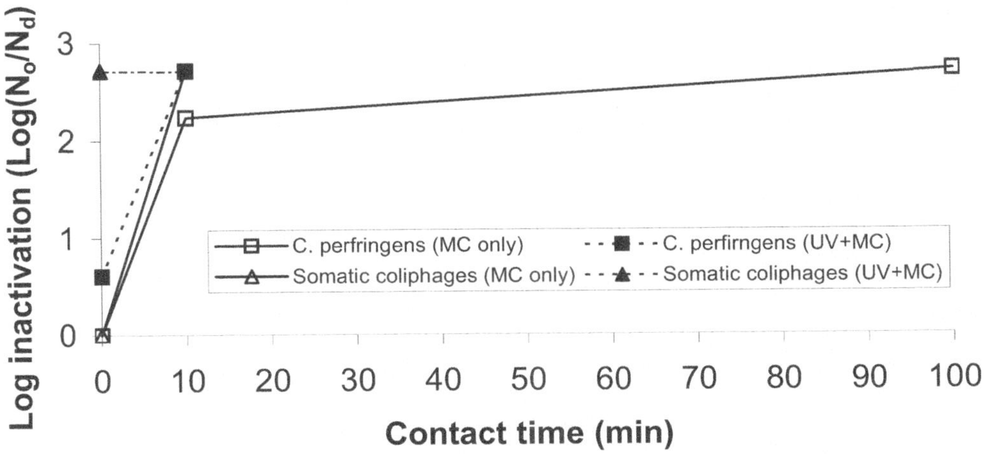

Figure 5-4. Inactivation of Microorganisms by UV and/or Monochloramine (Monochloramine: 10 mg/L [initial concentration] for 10 minutes) in Wastewater (Beaufort).

The data indicate that an appreciable (>3 $\log_{10}$) level of inactivation of somatic coliphage and *C. parvum* oocysts can be achieved by sequential application of FC followed by LP UV with reasonable dose ranges (CT values of <100 mg/L*min and intensity * time [IT] values <40 mJ/cm^2). However, *C. perfringens* spores are extremely resistant to this sequential disinfection scheme of FC followed by LP UV. The results also show that an appreciable (>3 $\log_{10}$) level of inactivation of somatic coliphages and *C. perfringens* spores can be achieved by sequential application of LP UV followed by monochloramine with reasonable dose ranges (CT values of <1000 mg/L*min and IT values <16 mJ/cm^2). Therefore, an approach that utilizes UV disinfection followed by FC and subsequent formation of monochloramine (due to ammonia in the water) with relatively long contact time (e.g., ~100 minutes) should be capable of achieving inactivation (>4 $\log_{10}$) of most microorganisms, including enteric viruses, bacteria, and protozoan parasites within a practical range of UV and free/monochloramine doses (<40 mJ/cm^2 and 10 mg/L to 20 mg/L, respectively).

5.5 UV and Chlorine Inactivation of PAC and NPAC

5.5.1 Introduction and Purpose

Typical disinfection of wastewater using UV or chlorination results in some residual of coliform organisms in the effluent. This is likely because the coliform is protected from the disinfectant due to its physical state and interaction with particles, as described above. This research was aimed at comparing the capabilities of UV and chlorine for inactivation of particle-associated coliform (PAC) and non-particle associated coliform (NPAC) in wastewater.

5.5.2 Materials and Methods

Samples containing NPAC and PAC were prepared as follows. NPAC were obtained using chemical extraction, filtration, and filtration combined with blending. Ethylenebis(oxyethylenenitrilo)tetraacetic acid (EGTA; Sigma Chemical Co., St. Louis, Missouri) was used in the chemical extraction method as suggested by previous studies (see Appendix A). For the first method (NPAC-1) EGTA was added to samples to give a final concentration of 1 mM, which was followed by 1 minute of blending at 3500 rpm. A seven-speed laboratory blender (Model No. 34BL97, Waring Products, New Hartford, Connecticut) was used to blend the samples. In the second method (NPAC-2), wastewater was filtered through a 5-μm filter paper (Whatman Inc., Clifton, New Jersey) to separate PAC from NPAC, assuming the majority of PAC are associated with particles >10 μm (Emerick et al., 1999) and could be held by the filter. In the third method (NPAC-3), wastewater was again filtered through the 5-μm filter, but this time the filtrate was blended for 1 minute at 3500 rpm to further minimize PAC remaining in the sample.

To prepare samples containing PAC, wastewater was initially filtered through a 5-μm nylon mesh, and the cake on the mesh was resuspended in wastewater supernatant filtered through 0.45-μm membrane filter (Millipore Corp., Bedford, Massachusetts). However, preliminary experiments showed that resuspension of the cake collected on the mesh inevitably damaged the structure of wastewater flocs, and most probably affected the location and attachment of the PAC incorporated within these flocs. Therefore, unaltered wastewater (secondary effluent) directly for samples containing PAC was used, since wastewater contains coliform that are naturally associated with particles based on previous research (see Appendix A, Section 4) and by others (e.g., Parker and Darby, 1995). The coliform in wastewater exist in both PAC and NPAC states, but NPAC are inactivated first during disinfection; thus, their presence would not preclude observation of the disinfection results for the PAC portion of the wastewater sample.

UV (Phase I) and chlorine (Phase II) disinfection were used to inactivate the NPAC and PAC in wastewater. In the Phase I UV disinfection experiments, a sample size of 20 mL was exposed to monochromatic UV light (LP mercury vapor lamp) at a wavelength of 254 nm in a petri dish. The sample was stirred continuously during the irradiation. The incident irradiation was measured using a radiometer with a germicidal UV detector (Model No. IL 1700 SED 240, International Light, Newbury Port, Massachusetts) calibrated at 254 nm to standards traceable to the National Institute of Standards and Technology. The average irradiance was calculated by correcting the incident irradiance for the spatial incident variation across the petri dish, the liquid absorbance of the sample, and the sample depth. Sample absorbance was measured using a Varian Cary 100 Bio UV-visible spectrophotometer (Varian Analytical Instruments, Walnut Creek, California). UV dose (mJ/cm^2) was calculated as the product of the average UV irradiance (mW/cm^2) multiplied by the exposure time(s), with UV doses ranging between 0 and 100 mJ/cm^2.

In Phase II, the chlorine disinfection experiments, wastewater samples having an initial total chlorine concentration of 1 mg/L, 5 mg/L, 10 mg/L, and 15 mg/L were prepared using fresh household bleach (5.25% sodium hypochlorite, Clorox). The contact times investigated were 5, 15, 30 and 45 minutes. FC and combined chlorine concentrations were determined using a combination of DPD ferrous titrimetric and DPD colorimetric methods (American Public Health Association, 1995). A 0.1% sodium thiosulfate (EM Science, Gibbstown, New Jersey) solution was used to neutralize the chlorine in samples following disinfection. Surviving coliform after the disinfection procedures were determined using membrane filter (MF) method (American Public Health Association, 1995). Average particle size and total particle number of the samples were determined using a particle size counter (Model No. WG5-260, Met One, Grants Pass, Oregon). Data points shown in the figures are the average of three replicates.

5.5.3 Results

In Phase I, samples containing NPAC and PAC were inactivated using UV disinfection at doses from 0 to 100 mJ/cm^2. As illustrated in Figure 5-5, PAC exhibited slower inactivation kinetics and tailing, where as the NPAC were more easily and rapidly inactivated to beyond detection limits. Even after relatively high doses of UV irradiation, complete inactivation of PAC could not be achieved. These results indicate the likelihood that some of the coliform were associated with particles to an extent that they were completely shielded from the UV light. The three samples that were treated to produce NPAC were easier to inactivate.

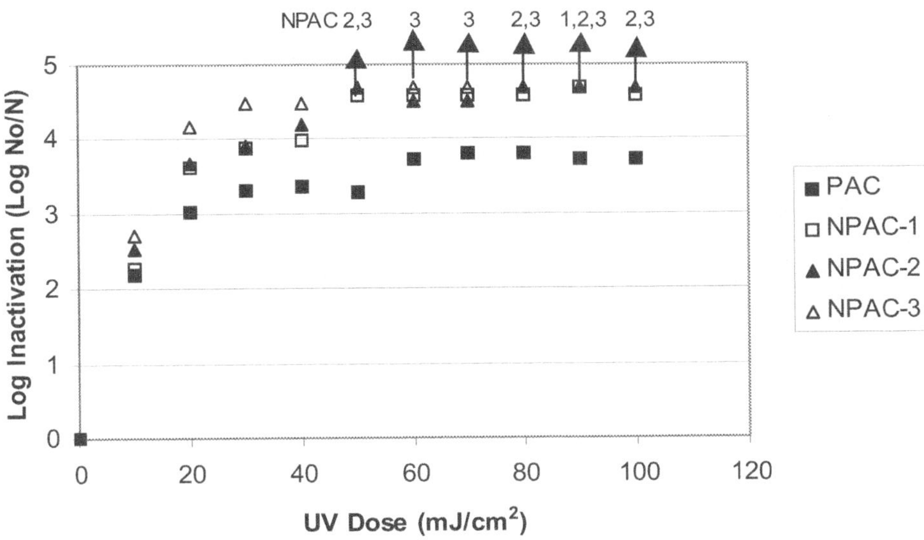

Figure 5-5. Dose-survival Relationship of PAC and NPAC Under UV Disinfection.

Complete inactivation of NPAC was achieved at a UV dose of 90 mJ/cm^2 for the first sample (NPAC-1), which was obtained by EGTA extraction. The other two samples containing NPAC (NPAC-2 and NPAC-3) were totally inactivated at a UV dose of 50 mJ/cm^2. When EGTA is introduced to wastewater, it binds with divalent cations that are an essential part of wastewater flocs. Divalent cations help to form bridges between negatively charged bacterial surfaces, extracellular polymers, and inorganic particles and hold the floc structure together. Extraction of divalent cations results in deflocculation and a decrease in average particle size (Eriksson and Alm, 1991; Bruus et al., 1992). Even though EGTA treatment increased the number of particles in smaller fractions and reduced the average particle size in NPAC-1, some of the remaining particles were still large in size (>10 µm) and may have been able to shield coliform (Table 5-5). Filtration, however, eliminated the majority of large particles in addition to other smaller particles, and greatly reduced the possibility of PAC presence in samples NPAC-2 and NPAC-3.

Table 5-5. Particle Size Distributions of the Wastewater Samples Containing PAC and NPAC.

Particle Size Range (μm/mL)	PAC	NPAC-1	NPAC-2	NPAC-3
2-5 μm	8645	11,495	4625	5385
5-10 μm	3265	4430	790	935
> 10 μm	1750	1865	205	245
Total	13,660	17,790	5,620	6565
SD (on total)	393	210	225	455

Figure 5-5 also illustrates that log inactivation of NPAC-1, NPAC-2, and NPAC-3 were greater than the log inactivation of PAC at any given UV dose, which indicates the disinfection effectiveness of UV on NPAC is significantly greater than on PAC. A 4-log reduction in the initial coliform count was achieved at 30 mJ/cm^2 for all of the samples containing NPAC. However, 4-log reduction could not be accomplished for the sample containing PAC even after a UV irradiation of 100 mJ/cm^2. The particle size distributions of the samples containing NPAC and PAC are presented in Table 5-5 in three different particle size ranges: 2 μm to 5 μm, 5 μm to 10 μm, and >10 μm. EGTA extraction (NPAC-1) increased and filtration (NPAC-2) decreased the total number of particles in all of the particle size ranges. When filtration was followed with blending (NPAC-3), an increase in the total number of particles in all of the particle size ranges was observed as compared to NPAC-2.

Some of the particle size data seemed to be inconsistent with what would be expected from the floc disruption protocols. For example, blending increased the number of particles that were >10 μm in both NPAC-1 and NPAC-3 when it was expected to decrease it. Possible explanations for this increase include the following. First, the anomaly may be inherent to the variability in the particle-size counting. Problems associated with reliability and sensitivity of particle-size counting in water and wastewater has been reported (Chowdhury et al., 2000). Although the results vary, overall trends in the data are as would be expected; thus, they do not point to major problems with the particle analyzer. Second, even though the samples may come from the same wastewater, they are likely to exhibit variability in the number of particles and the particle size distribution. Therefore, it is possible to observe some discrepancy between the results. Nonetheless, the overall results from the particle counts and disinfection trials indicated that when large particles and flocs were broken into smaller pieces or filtered out using physical and chemical techniques, the inactivation rate of the coliform increased and little tailing in the dose response curve occurred.

In the second part of this study, samples containing NPAC and PAC were inactivated using chlorine disinfection. Chlorine was added from a stock solution of freshly prepared household bleach directly to wastewater samples to give an initial total chlorine concentration of 1 mg/L, 5 mg/L, 10 mg/L, and 15 mg/L. The contact times used were 5, 15, 30, and 45 minutes. FC and combined chlorine concentrations remaining in the samples were measured throughout the experiment. The wastewater contained approximately 0.2 mg/L of ammonia. Figure 5-6 illustrates the decay of the total chlorine during the first 45 minutes for various starting concentrations of total chlorine in wastewater samples. Similar total chlorine decay kinetics were observed in all cases. All of the chlorine was completely consumed within 1 minute (first data point after zero) in the sample that contained 1 mg/L initial total chlorine. For the sample dosed with 5 mg/L, the combined chlorine concentration stabilized at approximately 0.3 mg/L, while the FC declined over time. At an initial dosing of 10 mg/L, the combined chlorine concentration stabilized at approximately 1.4 mg/L, while the FC declined over time. At an initial dosing of 15 mg/L, the combined chlorine concentration stabilized at approximately 1.7 mg/L while the FC declined over time. The average total chlorine loss (consumption) over the duration of the experiment was approximately 3.5 mg/L for the samples that had initial total chlorine concentrations of 5 mg/L, 10 mg/L, and 15 mg/L. Therefore, at initial chlorine concentrations at ≥5 mg/L, at least 1 mg/L (and often much more) of FC was present throughout the exposure.

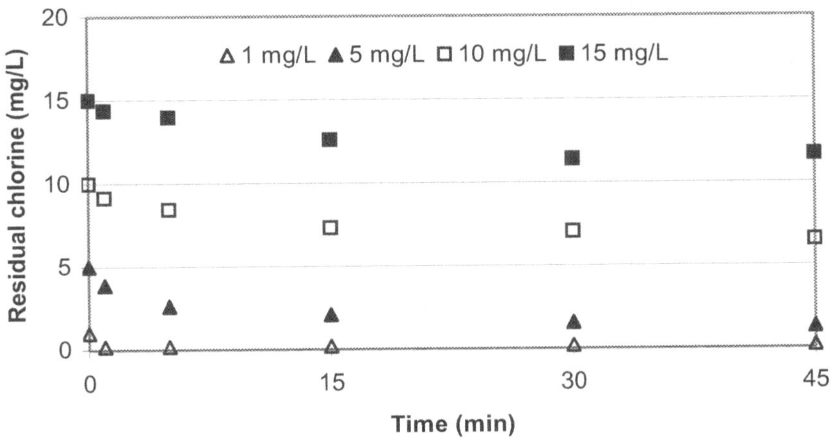

Figure 5-6. Residual Total Chlorine Concentrations Remaining in the Samples at the End of Each Contact Time During the Disinfection Experiment.

Figure 5-7 illustrates the inactivation of PAC- and NPAC-containing wastewater at chlorine CT values ranging from 0 to 525 mg-min/L. The initial total chlorine concentrations in the samples were 1 mg/L, 5 mg/L, 10 mg/L, and 15 mg/L, and the contact times were 5, 15, 30, and 45 minutes. Chlorine CT doses were calculated by integrating the chlorine residual concentration over the contact time. Compared to the PAC-containing wastewater, inactivation of the NPAC wastewater using chlorine disinfection was more rapid at doses below 150 mg-min/L. The lowest chlorine dose required to disinfect NPAC completely was 58 mg-min/L, which was accomplished at 5 mg/L initial chlorine and 45 minutes contact time. However, some of the NPAC were still viable after exposure to chlorine doses of 109 mg-min/L and 189 mg-min/L when the contact times were 15 minutes with chlorine dosing of 10 mg/L and 15 mg/L respectively (Figure 5-8). Some of the PAC were also less susceptible to disinfection, and survived after exposure to a chlorine dose of 342 mg-min/L at 15 mg/L initial chlorine and 30 minutes contact time. However, total inactivation of PAC was achieved at a lower chlorine dose of 293 mg-min/L when the initial chlorine was 10 mg/L and the contact time was 45 minutes. Overall, total disinfection of PAC could only be achieved when the contact time was 45 minutes and the initial chlorine concentrations were relatively high (10 mg/L and 15 mg/L). Contact times <45 minutes resulted in the survival of PAC regardless of the initial chlorine concentration, despite the availability of ample FC (>1 mg/L) (Figure 5-8).

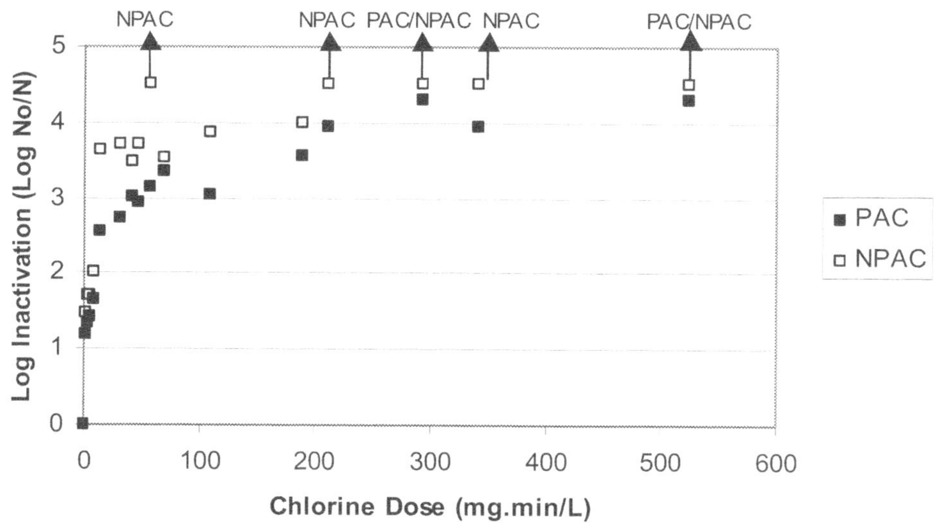

Figure 5-7. Dose-survival Relationship of PAC and NPAC Under Chlorine Disinfection

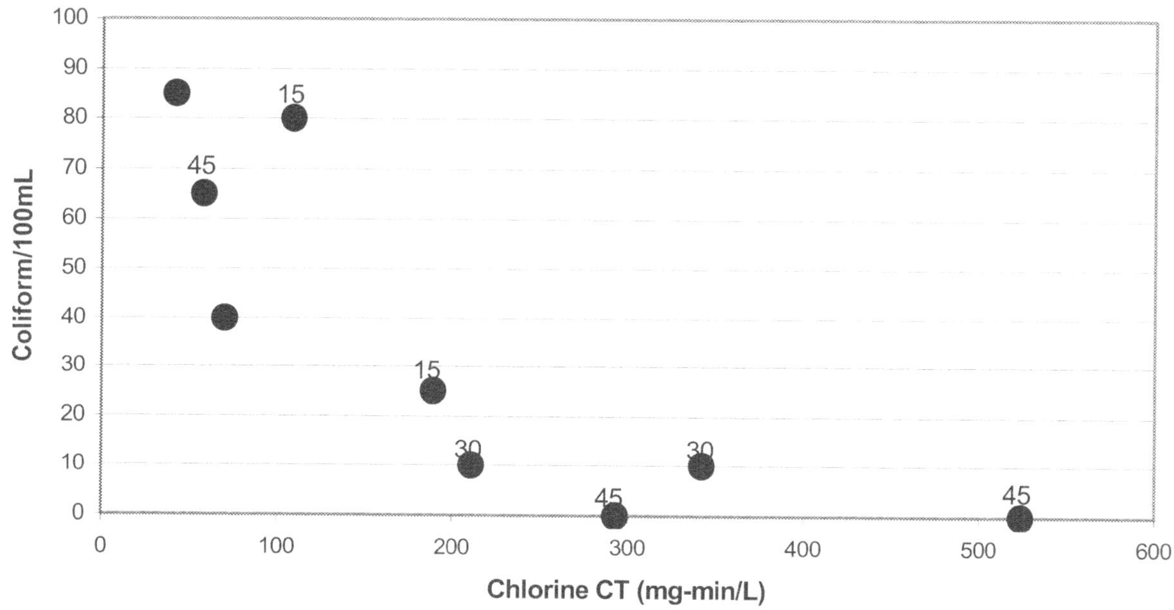

Figure 5-8. Impact of Contact Time on Inactivation of Coliform in Wastewater with Chlorine at a Given CT Dose.

These results indicate that contact time plays an important role in determining the effectiveness of chlorine disinfection in wastewater, and chlorine dose alone may not be a good indicator of disinfection effectiveness. Wastewater contains relatively high concentrations of soluble organics, inorganics, and suspended solids that may reduce the disinfection efficiency of chlorine by creating a chlorine demand or forming species of chlorine that are not very bactericidal. Diffusion of the chlorine into the floc to the target cell appears to be a rate-limiting step for chlorine disinfection. The data suggest that at practical chlorine doses, initial chlorine concentration does not affect the diffusion rate. However, provided enough time, chlorine appears to be effective against particle associated microbes.

In this study, much of the chlorine was in the FC form. With wastewater containing high levels of ammonia or greater chlorine demand, the effectiveness of chlorine may be affected because combined chlorine will be less reactive. The lower reactivity may allow the chloramines to last longer and diffuse into the particle over time, resulting in greater disinfection. It may be too slow and less reactive than needed to inactivate some chloramine resistant pathogens, such as viruses.

CHAPTER 6.0

Conclusions and Implications for Wastewater Utilities

This research covered many aspects of wastewater disinfection with UV light and chlorine. The findings can be directly utilized by wastewater utilities to optimizing existing disinfection practice and make informed choices about upgrades to the disinfection process at a given utility, and thus will be of value to the wastewater industry. Following is a summary of the major conclusions from this research.

1. In properly sized UV disinfection systems accounting for UV absorbance of the wastewater, low doses of UV radiation from either LP or MP UV lamps are effective for inactivating *C. parvum* oocysts in wastewater effluent.
2. Repair of the DNA damage during UV inactivation of *C. parvum* oocysts by reactivation processes does not appear to be an issue of concern.
3. The use of chlorine and UV irradiation for disinfection does not generate the presence of infective, but nonculturable bacterial pathogens in adequately disinfected wastewater. Thus, the presence of VBNC bacteria capable of resuscitation in wastewater following disinfection is not a concern.
4. There is no simple relationship between the type and size of viral nucleic acids and the relative susceptibility of viruses in wastewater to UV inactivation. Because viruses with similar nucleic acid types and sizes showed great differences in response to UV radiation, it is likely that other virions properties, including virion size, shape, capsid morphology, and organization also influence response to UV radiation. Overall, most of the viruses tested could be disinfected using typically applied doses of UV irradiation. However, adenovirus was found to be very resistant to UV at typical doses. Further studies with other enteric viruses possessing linear DS-DNA, besides adenoviruses, would need to be performed to comprehensively examine the relationships between virion and nucleic acid properties and their responses to UV radiation.
5. UV wavelengths emitted from LP and MP lamps were equally effective for inactivation of *C. parvum* oocysts. For MS2 coliphage, there may be some advantages to wavelengths below 220 nm and between 260 nm and 270 nm, over the effectiveness of 254 nm emitted by LP UV sources. The benefit may extend to other viruses, but this was not tested. However, absorbance of the wastewater would make wavelengths below 240 nm unavailable in typical UV reactors.

6. Despite the presence of plasmids in bacteria that confer antibiotic resistance, it was not possible to detect any differences in the inactivation of these bacteria compared to those without the plasmid conferring antibiotic resistance, by UV irradiation.
7. Methods were developed to optimize the study of particle-associated microorganisms using physical particle disruption methods and novel molecular techniques.
8. The presence of PAC bacteria was confirmed using physical disruption methods, while the presence of particle associated pathogens, such as *C. parvum* and *S. typhimurium*, in wastewater was confirmed using molecular methods.
9. The relationship between time and dose in chemical disinfection with chlorine/chloramine is important, specifically in relationship to particle penetration. At a given chlorine dose, longer chlorine contact times were more effective for inactivation of particle-associated microbes.
10. UV disinfection was not as effective as chlorine over extended contact times for inactivation of PAC.
11. Sequential disinfection using FC, dynamically converting to chloramines in-situ due to the presence of ammonia in wastewater, was effective for inactivation of many waterborne pathogens, including viruses. However, this was not effective for inactivation of *C. parvum* oocysts.
12. Unlike sequential disinfection with FC and LP UV, inactivation of *C. perfringens* spores by monochloramine was effective using 10 mg/L of monochloramine and 100-minute contact time, indicating a disinfection benefit for extended contact time with combined chlorine species.
13. Sequential disinfection using some form of FC and combined chlorine and UV irradiation was effective for inactivation of all types of microbes tested, including pathogens. However, some contact time with FC is necessary for proper virus disinfection. If particle association is significant for *C. parvum* such that it is protected from UV irradiation, disinfection will not be effective for these associated pathogens.

Other important considerations that emerged from this research related to the integration of disinfection and water quality. If a dual disinfection system, such as UV and chlorine (or chloramine), is utilized, both the UV portion of the disinfection scheme and the chlorination step would be affected by the water quality and presence of particles. For chlorine, the presence of ammonia and other species causing chlorine demand is important to determine the ratios of free and combined chlorine present and thus the effectiveness of disinfection. When UV disinfection is used, the presence of particles will dictate the level of filtration needed to insure complete and thorough inactivation of pathogens.

Depending upon the individual utility situation, consideration of a number of scenarios relating to the research described in this report could be of value. For instance, a wastewater utility that is discharging to a sensitive waterbody or one that is used as a drinking water intake, may want to employ a sequential UV-chlorine/chloramine disinfection scheme to ensure inactivation of viruses and protozoan pathogens. A utility that is experiencing problems with meeting effluent standards may need to investigate the issue of particle-associated microbes and adjust their upstream processes to minimize formation of microbes associated with particles, or add filtration. The use of FC/chloramines sequential disinfection in a dynamic chloramination mode by a WWTP could enhance the overall removal of bacteria, viruses, and spores over extended contact times; however, this scenario is not capable of disinfecting *Cryptosporidium*. For some wastewater utilities that are particularly susceptible to *C. parvum*, the addition of a UV disinfection step would be necessary to protect downstream users from *Cryptosporidium* exposure.

APPENDIX A

PARTICLE ASSOCIATION OF COLIFORM IN WASTEWATER

A.1 Introduction and Purpose

Disinfection of microorganisms associated with particles is a growing concern. Microorganisms that are embedded in flocs or surrounded by particles, have a higher chance of surviving the disinfection process since they are effectively shielded by particles and protected from disinfectants. (LeChevallier et al., 1981; LeChevallier et al., 1984; Berman et al., 1988; Parker and Darby, 1995; Emerick et al., 1999). Coliform bacteria, the most commonly used indicator of disinfection performance, are much less resistant to chemical disinfection compared to some other health-related microorganisms such as *Cryptosporidium parvum* and *Giardia lamblia* and less resistant to UV disinfection compared to viruses. Therefore, the survival of particle-associated coliform (PAC) through a disinfection process will likely indicate the survival of other more resistant pathogens that may also be particle-associated.

Interaction of microbes with wastewater may act to protect the microbes from receiving a lethal dose of disinfectant, even though sufficient disinfectant may be present in the bulk water. Wastewater samples were collected and evaluated to determine the extent of particle association of indigenous coliform and the subsequent impact that particle association has on inactivation with chlorine and ultraviolet (UV) irradiation (described in the next section). Physical and chemical methods for determining the extent of particle association are reported here.

A.2 Materials and Methods

Physical and chemical extraction methods were used to detect presence of coliform in wastewater particulates. Physical extraction methods used were blending, stirring, sonication, centrifugation, and combination of blending and centrifugation. Blending experiments were conducted using a seven-speed laboratory blender. (Model No. 34BL97, Waring Products, New Hartford, Connecticut). The blending speeds tested were 3500, 7000, 11,500, 14,500, 17,000, 19,000, and 22,000 rpm, and the blending times ranged from 0.5 to 2 minutes. A jar test gang stirrer (Model No. 7790-100, Phipps and Bird, Richmond, Virginia) was used for low-speed stirring (0 to 300 rpm), and a magnetic stirrer was used for high-speed stirring (100 to 1000 rpm) (Fisher Scientific, Pittsburgh, Pennsylvania). Sorvall Super T21 (Kendro Laboratory Products, Newton, Connecticut) was used to centrifuge the samples at centrifugation speeds between 500 and 3000 rpm. Branson 1200 sonicator (Branson Ultrasonic, Danbury, Connecticut) was used to sonicate the samples between 15 and 300 seconds. Ethylenediaminetetraacetic acid (EDTA; Fisher Scientific, Pittsburgh, Pennsylvania) and ethylenebis(oxyethyleneitrilo)tetraacetic acid

(EGTA; Sigma Chemical Co., St. Louis, Missouri) were used in the chemical extraction of PAC. After the addition of EDTA and EGTA, samples were stirred at 1000 rpm for 3 minutes or blended at 3500 rpm for 1 minute, and the results were compared to a similar method suggested by Camper et al. (1985). This method consisted of blending the samples at 16,000 rpm in a solution of 10^{-6} M Zwittergent 3-12 (Calbiochem Corp., La Jolla, California), 10^{-3} M EGTA, 0.1% peptone (ICN Biomedicals, Inc., Costa Mesa, California), and 1 mM Tris (ICN Biomedicals, Inc.,Costa Mesa, California) buffer for 3 minutes at 4°C. In addition, the effectiveness of Camper et al.'s chemicals was tested at 3500 rpm blending for 1 minute. The effectiveness of the extraction methods were evaluated by quantifying the total coliform density in the samples using the membrane filter method (American Public Health Association, 1995) before and after the extraction of PAC. Average particle size and total particle number for the samples was determined using a particle size counter (Model No. WGS-260, Met One, Grants Pass, Oregon). The data points in the figures show the average of three replicates, and the error bars show the 95% confidence interval.

A.3 Results and Discussion

The physical extraction methods used in this study were blending, stirring, sonication, centrifugation, and combination of blending and centrifugation. The chemical extraction methods used were EDTA and EGTA extraction and a protocol that combined Zwittergent 3-12, Tris buffer, EGTA, and peptone.

A.3.1 Blending

Speeds from 3500 rpm up to 22,000 rpm were used to blend the samples at 4°C for 30 seconds (Figure A.1a). Temperature was not controlled during blending. Since the blending time was short, the increase in the temperature of the samples was not significant enough to affect the results. For example, blending at 3500 rpm for 30 seconds increased the sample temperature by 0.6°C, and blending at 22,000 rpm for 30 seconds increased the sample temperature by 1.7°C. Right after blending, the samples were placed in a cooler at 4°C where the slightly elevated temperatures were brought down to 4°C. In comparison to other soil, granular activated carbon, and wastewater studies that have used only high-speed blending or homogenization (usually between 17,000 rpm and 21,000 rpm) (LeChevallier et al., 1981; Faegri et al., 1977; Bakken, 1985; Camper et al., 1985; Parker and Darby, 1995), the results from the blending experiments were unexpected. The highest coliform density was measured at the lowest blending speed at 3500 rpm with an increase of 40.9% in the coliform density over the untreated sample (Figure A.1a). When the blending speed was increased, total coliform density decreased accordingly. The lowest coliform count was measured at the highest blending speed of 22,000 rpm, and the coliform density was 17.4% lower compared to the untreated sample. These results indicate that low-speed blending appears sufficient to breakup the wastewater flocs and aggregates of bacteria and release PAC from the particulates. Figure A.2 illustrates the change in the average particle size and total number of particles in samples after blending at different speeds for 30 seconds. The volume of the sample in the blender was 700 mL at each blending speed. As expected, increasing the blending speed decreased the average particle size and increased the total number of particles in the blended samples, which indicated that blending was effective in breaking up wastewater particles and flocs. At blending speeds higher than 17,000 rpm however, the total number of particles in the sample dropped slightly probably due to the detection limitation of the particle size analyzer (>2μm particles).

The decrease in the coliform density with increasing blending speed may have been due to the disturbance of cells at high blending speeds. To check whether blending adversely affects the cells and decreases the coliform count, wastewater was filtered through a 3-μm membrane filter. After the filtration, it was assumed that no PAC or aggregates of coliform were left in the sample, and therefore blending would not increase the coliform density by breaking up the flocs or bacterial aggregates. A decrease in the coliform count after blending would then indicate the occurrence of cell disturbance. As expected, the coliform density decreased by 32.3% after blending at 3500 rpm for 1 minute. Between 3500 rpm and 19,000 rpm the coliform count did not change significantly, but increasing the blending speed to 22,000 rpm lowered the coliform count another 34.4% (Figure A.1b). These results indicate that in wastewater, blending may disturb the cells even at low speeds; therefore high-speed blending should be avoided in the extraction procedures. However, no experiments were performed to verify that cell structure was actually disturbed by these protocols.

Once optimum blending speed was determined as 3500 rpm in wastewater containing PAC, the effect of blending time in the extraction effectiveness of PAC was evaluated. Samples at 4°C were blended at 3500 rpm for 0.5, 1, 1.5, and 2 minutes (Figure A.1c). The temperature increase in the sample due to blending was 0.4°C after 0.5 minutes of blending, and 1.3°C after 2 minutes of blending. The highest coliform density was measured at 1 minute with an increase of 35.4% over the untreated sample, and longer blending times decreased the coliform density. These results indicate that cells may adversely be affected not only during high-speed blending but also during prolonged low-speed blending.

A.3.2 Stirring

Since the highest coliform density was recorded at the lowest setting of the blender (3500 rpm), it was necessary to test whether lower blending or stirring speeds would increase the coliform density further. First, a jar stirrer was used to stir the samples at 4°C at speeds between 0 and 300 rpm for 5 minutes at room temperature. When stirring speed was <200 rpm, a decrease in the coliform density was observed which was attributed to the coagulation/flocculation process taking place at these low speeds (Figure A.3a). Increasing the stirring speed further to 300 rpm resulted in an increase in the coliform density. Subsequently, samples at 4°C were stirred at higher speeds between 0 and 1000 rpm for 3 minutes at room temperature using a magnetic stirrer. Stirring the sample for 3 minutes at room temperature increased the sample temperature by approximately 1.5°C. Below 800 rpm there was not a significant increase in the coliform density. Once the speed was increased to 1000 rpm, coliform density increased by 103.4% over the control (Figure A.3b). This was probably the speed at which wastewater flocs started to break apart.

The wastewater used in the stirring experiments was collected one day after a big storm hit Durham, NC, and the coliform densities were lower than the wastewater samples used in other experiments. It is possible that the rain might have diluted the wastewater, or the storm might have distressed the treatment process and changed the coliform density.

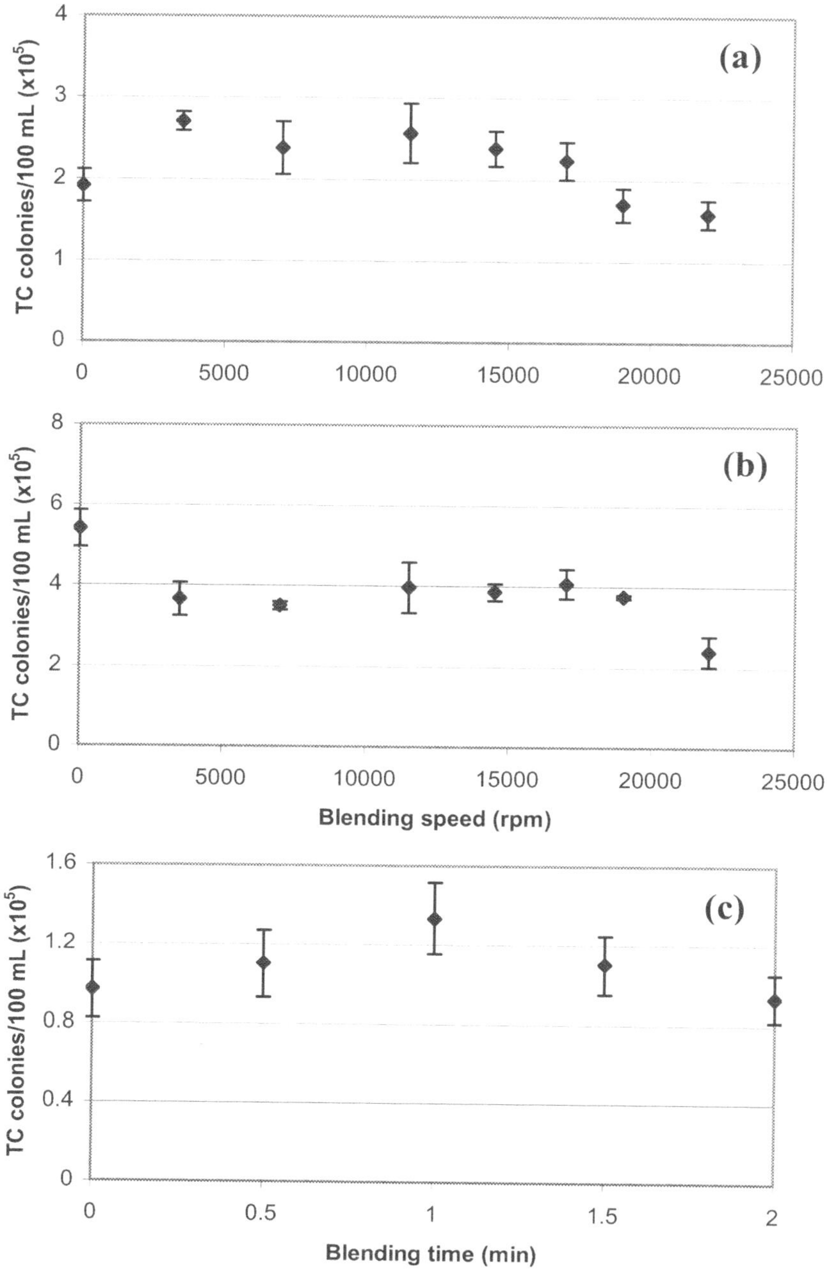

Figures A-1. Effect of (a) Blending Speed on Total Coliform Density in Wastewater; (b) Blending Speed on Total Coliform Density in Filtered Wastewater; (c) Blending Time (at 3500 rpm) on Total Coliform Density in Wastewater.

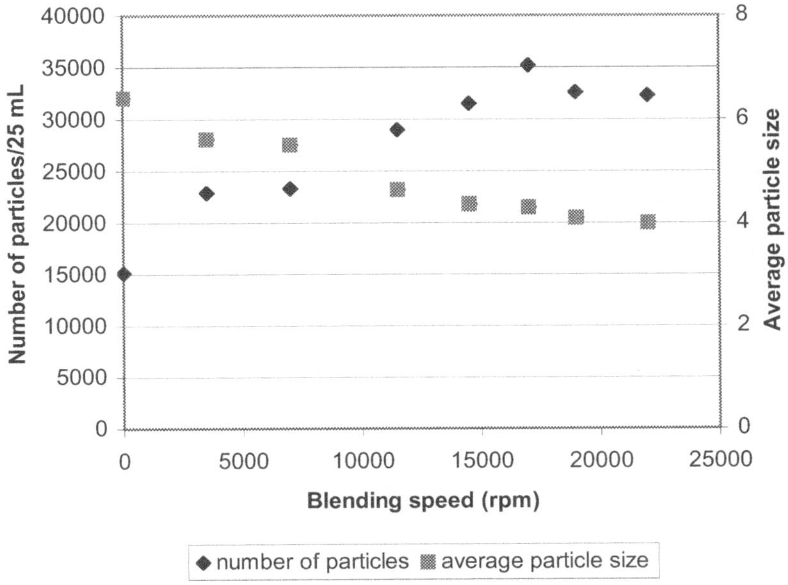

Figure A-2. Total Number of Particles and Average Particle Size in Blended Samples.

A.3.3 Sonication

Sonication has also been used in the extraction of bacteria from particles, but the results have been contradictory. Many researchers reported increased bacterial counts by sonication (Stevenson, 1958; Zvyagintsev and Galkina, 1967; Oliver and Cosgrove, 1975; Weise and Rheinheimer, 1979; Ramsay, 1984), but extended sonication resulted in cell disruption (Zvyagintsev and Galkina, 1967; Weise and Rheinheimer, 1979; Ramsay, 1984). Dale (1974) preferred blending to sonication for extracting bacteria from sand, whereas Ramsay (1984) reported sonication released more bacteria from soil than blending or shaking. Parker and Darby (1995) reported no significant increase in total coliform densities after sonication. In this study, wastewater samples at 4°C were sonicated from 15 to 300 seconds at room temperature, and no significant increase was observed in the total coliform density (Figure A.4), similar to the findings of Parker and Darby (1995). The sonicator used had an ultrasound frequency of 47 kHz and a power input of 42 W/L. The disagreement between the results from different studies might be due to a number of factors including differences in ultrasonic energy, exposure time, solution conditions (i.e., turbidity, temperature), and variable microorganism resistance. While some bacteria attached to particles may be released by ultrasonic treatment, some will inevitably be damaged. Therefore, depending on the conditions and ultrasonic energy used, sonication might increase, decrease, or not influence the density of the particle-associated microorganisms in samples.

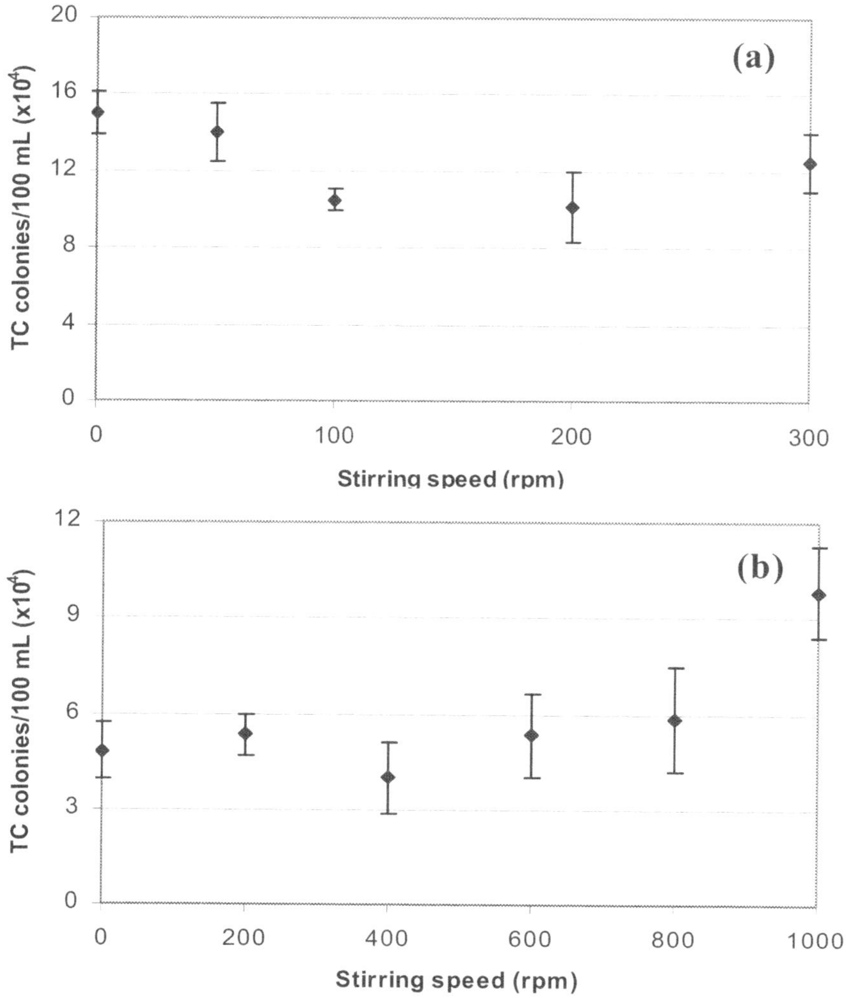

Figure A-3. Effect of (a) Low stirring Speeds on Total Coliform Density; (b) High Stirring speeds on Total Coliform Density.

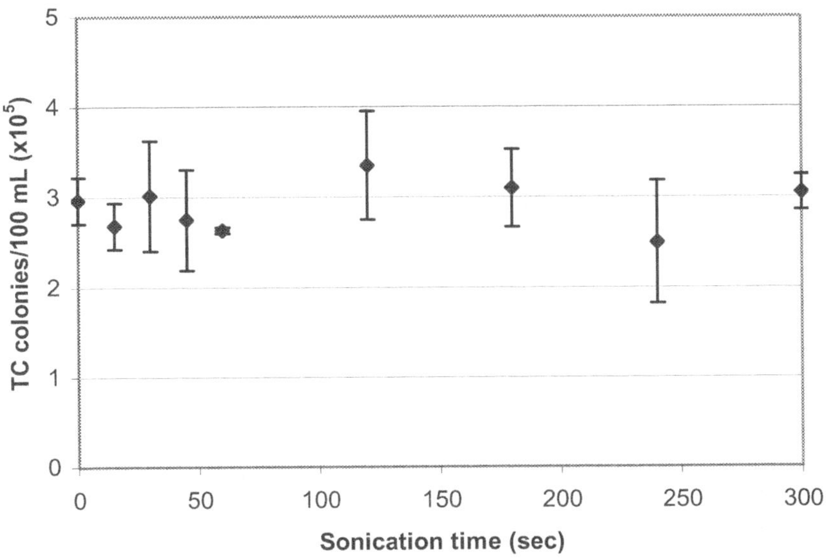

Figure A-4. Effect of Sonication Time on Total Coliform Density. The Data Points Show the Average of Three Replicates, and the Error Bars Show the 95% Confidence Interval.

A.3.4 Centrifugation

Even though repeated blending and low-speed centrifugation has been successfully used to separate bacteria from larger soil particles (Faegri et al., 1977; Bakken, 1985), this method did not work well to extract PAC from wastewater particulates in these experiments. In this study, a temperature-controlled centrifuge was used to centrifuge the wastewater samples at 4°C. Blending at 3500 rpm for 1 minute alone increased the coliform density, but when it was followed by centrifugation at 3000 rpm (1075 g) for 5 minutes, a decrease in the coliform count was observed. Combination of stirring at 1000 rpm for 3 minutes and centrifugation at 3000 rpm for 5 minutes did not increase the coliform density either (Figure A.5a). Finally, centrifugation alone at various speeds for 5 minutes was tested as an extraction method, and a significant decrease in the coliform densities even at very low centrifugation speeds was observed (16.9% at 1000 rpm [120 g] and 28.4% at 3000 rpm [1075g]) (Figure A.5b). The decrease observed in the coliform density may be due to the removal of some of the PAC that settle with the sediment during centrifugation. As a result, low-speed centrifugation was ruled out as an extraction method.

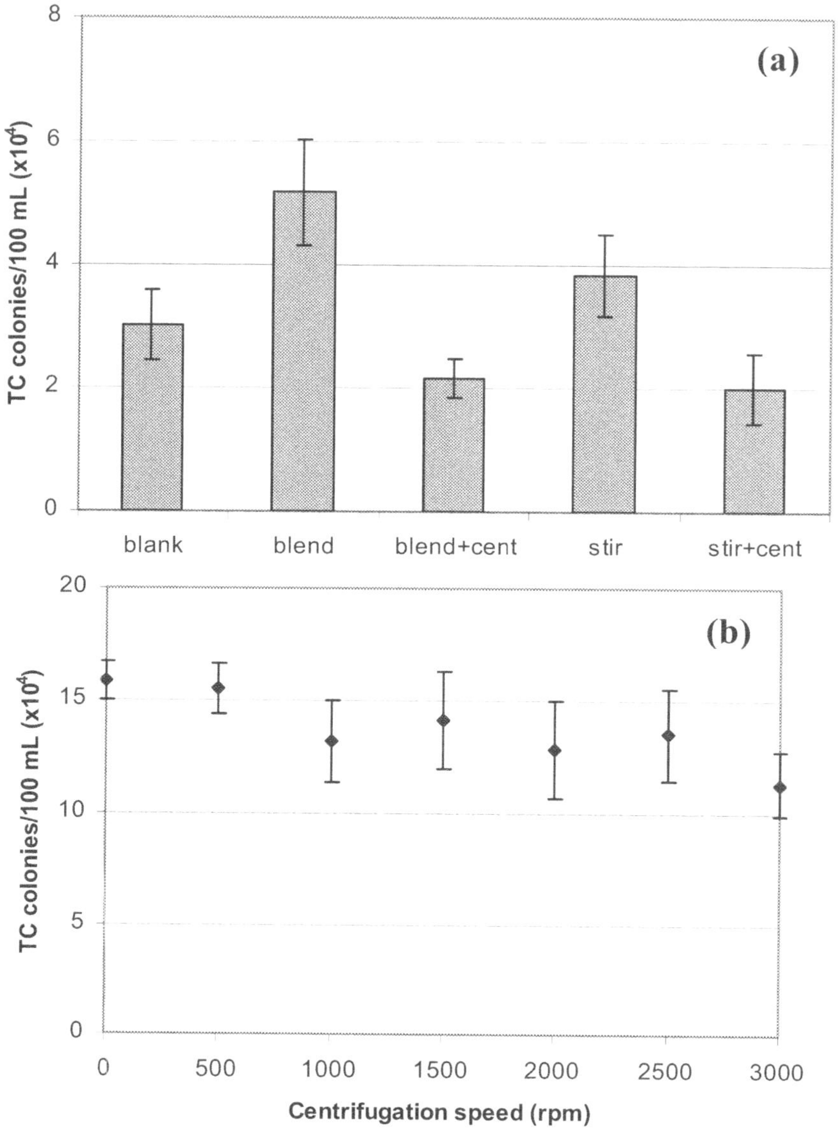

Figure A-5. Effect of (a) Centrifugation with Blending and Centrifugation with Stirring on Total Coliform Density; (b) Centrifugation Alone on Total Coliform Density.

A.3.5 Chemical Assisted Extraction

EDTA and EGTA were used in the chemical extraction of PAC from wastewater particulates. EDTA and EGTA are chelants used in the extraction of divalent cations, and their addition was expected to increase the coliform densities in samples. EDTA and EGTA were added to samples to give a final concentration of 1 mM to 5 mM. Addition of EDTA and EGTA did not change the pH of the samples. After the addition of EDTA and EGTA, samples were stirred at 1000 rpm for 1 minute. However, no significant increase in the coliform densities was measured (data not shown). In another trial, the samples with EDTA and EGTA were kept in the refrigerator overnight to allow sufficient time for the chelation reaction to take place, but this did not increase the coliform density either (data not shown). Finally, EDTA and EGTA addition were combined with 1 minute of low-speed blending at 3500 rpm. Samples that contained 1 mM EDTA showed a 57.8% increase, and samples that contained 1 mM EGTA showed a 59.7%

increase in their coliform densities compared to the samples that contained no EDTA or EGTA but were blended for 1 minute at 3500 rpm (Figure A.6). Increasing EDTA and EGTA concentration above 1 mM did not create a significant additional increase in the coliform densities, so 1 mM was chosen as the EDTA and EGTA dose in chemical extraction of PAC.

Among various methods tested in this study, blending, stirring, EDTA extraction, and EGTA extraction were concluded to be successful in extracting PAC. The effectiveness of these methods was compared to a chemical protocol that consisted of blending the samples at 16,000 rpm in a solution of 10^{-6} M Zwittergent 3-12, 10^{-3} M EGTA, 0.1% peptone, and 1 mM Tris buffer for 3 minutes loosely modeled after a protocol suggested by Camper et al. (1986). In addition, the effectiveness of these chemicals was tested at a blending speed of 3500 rpm for 1 minute.

The results showed that EDTA and EGTA extraction at 3500 rpm were more effective than stirring, blending, and the protocol (indicated as "mixed" in Figure A.7) that consisted of blending the sample at 16,000 rpm with Zwittergent 3-12, Tris buffer, EGTA, and peptone (Figure A.7). The high blending speed (16,000 rpm) used in this protocol may create shear forces that are too harsh for wastewater flocs and microorganisms and thus decrease the coliform count in the samples. Therefore, this protocol was repeated at a blending speed of 3500 rpm for 1 minute instead of the 16,000 rpm for 3 minutes, and coliform densities were compared (Figure A.7). The coliform density at 3500 rpm was 34% higher than at 16,000 rpm, indicating cells may be disturbed at high blending speeds, resulting in lower counts. However, actual cell disturbance was not monitored. The effectiveness of the modified protocol at 3500 rpm was very similar to the effectiveness of the EDTA and EGTA extraction (approximately 115% increase in the coliform densities over the untreated sample for all three methods), suggesting any of these methods can be used effectively in the extraction of PAC from wastewater particulates. These results indicate that the use of chemicals combined with blending enhances the extraction effectiveness compared to blending or stirring alone. The average particle size and average total number of particles in these samples from three measurements are illustrated in Table 5-5. Blended samples have higher number of particles and smaller average size compared to the stirred samples. Addition of chemicals does not seem to significantly increase or decrease the total number of particles or the average size.

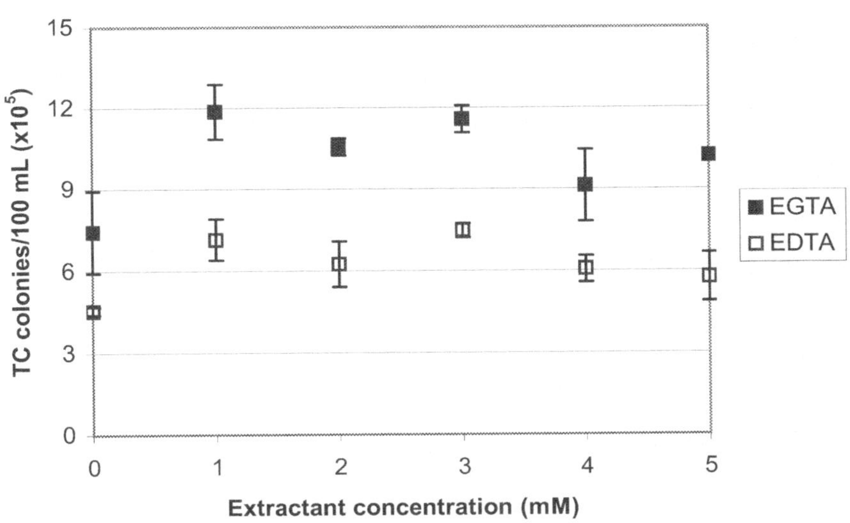

Figure A-6. Effect of EGTA and EDTA Addition on Total Coliform Density. The Data Points Show the Average of Three Replicates, and the Error Bars Show the 95% Confidence Interval.

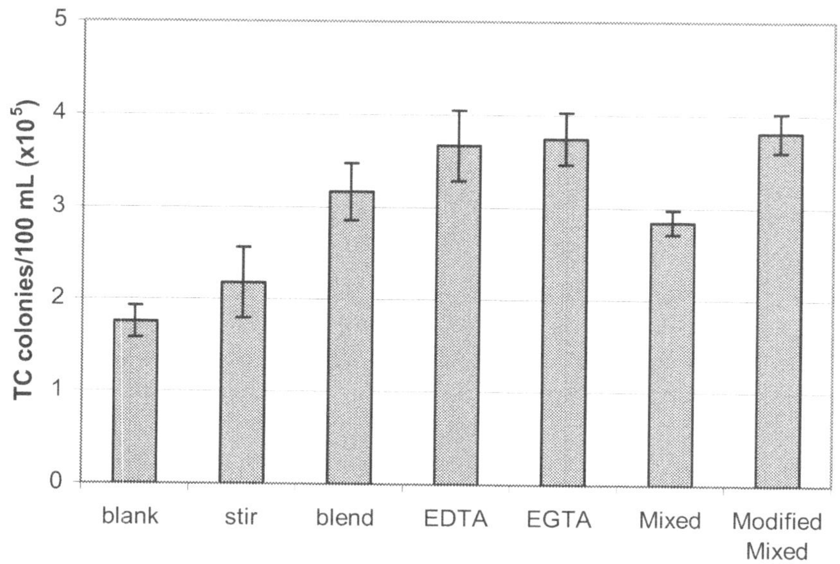

Figure A-7. Comparison of Optimized Physical and Chemical Methods for Coliform Extraction. The Data Points Show the Average of Three Replicates, and the Error Bars Show the 95% Confidence Interval.

APPENDIX B

LISTING OF PUBLICATIONS AND PRESENTATIONS RESULTING FROM THIS RESEARCH

Publications in Peer-Reviewed Journals:

Linden K.G.; Shin G.; Sobsey M.D. 2001. Comparative effectiveness of UV wavelengths for the inactivation of *Cryptosporidium parvum* oocysts in water. *Wat. Sci. Technol.* 43 (12):171-174.

Shin G.; Linden K.G.; Arrowood M.J.; Sobsey M.D. 2001. Low pressure UV inactivation and DNA repair potential of Cryptosporidium parvum oocysts. *Appl. Environ. Microbiol.* 67:3029-3032.

Ormeci B.; Linden K.G. 2002. "Comparison of UV and chlorine inactivation of particle and non-particle associated coliform", *Wat. Sci. Technol.: Wat Sup.* 2(5-6)

Oliver J.D.; Dagher M.; Linden K.G. (Submitted 2003) "Induction of *Escherichia coli* and *Salmonella typhimurium* into the viable but nonculturable state following chlorination of wastewater", *Journal of Water and Health*, IWA.

Oral Presentations and Proceedings:

Shin G.; Linden K.G.; Sobsey M.D. 1999. Inactivation of *Cryptosporidium parvum* oocysts with low pressure UV radiation. *Proceedings of Water Quality Technology Conference*. AWWA. Tampa, FL.

Shin G.; Faubert G.; Linden K.G.; Sobsey M.D. 2000. Low-pressure UV inactivation of *Cryptosporidium parvum* and *Giardia lamblia* based on infectivity assays and DNA repair of UV-irradiated *Cryptosporidium parvum* oocysts. *Proceedings of Water Quality Technology Conference*. AWWA. Salt Lake City, UT.

Shin G.; Linden K.G.; Sobsey M.D. 2000. Comparative inactivation of *Cryptosporidium parvum* oocysts and coliphage MS-2 by monochromatic UV radiation. *Proceedings of Disinfection 2000.* WEF. New Orleans, LA.

Shin G.; Linden K.G.; Faubert G.; Arrowood M.; Sobsey M.D. 2000. Inactivation of *Cryptosporidium parvum* oocysts and *Giardia lamblia* cysts by monochromatic UV radiation. *The First World Congress of International Water Association*. Paris, France.

Linden K.G.; Shin G.; Sobsey M.D. "Relative efficacy of UV wavelengths for the inactivation of *Cryptosporidium parvum*" Proceedings, *Disinfection 2000, WEF Specialty Conference*, New Orleans, LA. March 15-17, 2000.

Linden, K.G.; Shin G.; Sobsey M.D. 2000. Comparison of monochromatic and polychromatic UV light for disinfection efficacy. *Proceedings of Water Quality Technology Conference.* AWWA. Salt Lake City, UT.

Linden, K.G.; Shin G.; Sobsey M.D. 2000. "Comparative effectiveness of UV wavelengths for the inactivation of *Cryptosporidium parvum* in water" *International Water Association First World Congress*, Paris France, July 5-8, 2000. Poster.

Shin G.; Linden K.G; Sobsey M.D. 2001. "Low-Pressure UV Inactivation of Pathogenic Enteric Viruses and Bacteriophages", *Proceedings, AWWA Water Quality Technology Conference*, Nashville, TN, November 11-14, 2001

Shin, G.; Arrowood M.; Faubert G.; Linden K.G.; Sobsey M.D. 2001."DNA repair of UV-irradiated *Cryptosporidium parvum* oocysts and *Giardia lamblia* cysts" *Proceedings of the First International Congress on UV Technologies*, International Ultraviolet Association Washington DC, June 14-16, 2001.

Linden K.G.; Ormeci B. 2002. "Comparative Effectiveness of UV and Chlorine for Inactivation of Particle Associated Coliform", *Proceedings, WEF Disinfection 2002*, St. Petersburg FL, February 17-20, 2002

Linden K.G.; Ormeci B. 2002. "Comparison of UV and Chlorine Inactivation of Particle and Non-Particle-Associated Coliform", *Proceedings, IWA Enviro 2002 – World Water Congress*, Melbourne, Australia, April 7-12, 2002.

Shin G.; Ishida G.; Linden K.G.; Sobsey M.D. 2002 "Inactivation of Adenovirus, *Cryptosporidium parvum* oocysts and *Salmonella typhimurium* by sequential disinfection with UV radiation and free or combined chlorine" Proceed*ings, AWWA Water Quality Technology Conference*, Seattle, WA, November 10-14, 2002.

REFERENCES

Alvarez M.E.; O'Brien R.T. 1982. Effects of chlorine concentration on the structure of poliovirus. *Appl. Environ. Microbiol.*, 43:237-239.

Amann R. I.; Ludwig W.; Schleifer K.H. 1995. Phylogenetic identification and in situ detection of individual microbial cells without cultivation. *Microbiol. Rev.* 59, 143-169.

American Public Health Association, American Water Works Association, and the Water Environment Federation 1995. *Standard Methods for the Examination of Water and Wastewater.* 19th Ed., Washington, D.C.

Bakken L.R. 1985. Separation and purification of bacteria from soil. *Appl. Environ. Microbiol.*, **49**(6), 1482-1487.

Barcina I.; González J.M.; Iriberri J.; Egea L. 1989. Effect of visible light on progressive dormancy of *Escherichia coli* cells during the survival process in natural fresh water. *Appl. Environ. Microbiol.* 55:246-251.

Barer M.R. 1997. Viable but non-culturable and dormant bacteria: time to resolve an oxymoron and a misnomer? *J. Med. Microbiol.* 46:629-631.

Barer M.R.; Harwood C.R. 1999. Bacterial viability and culturability. *Adv. Micro. Physiol.* 41:93-137.

Barrett T. 1998. An investigation into the molecular basis of the viable but non-culturable response in bacteria. Ph.D. thesis, Univ. Aberdeen.

Battigelli D.; Lobe D.; Sobsey M.D.1993. Inactivation of hepatitis A virus and other enteric viruses in water by ultraviolet light. *Water Science and Technology*, 27 (3-4):339-342.

Battigelli D.A.; Sobsey M.D. 1994. The inactivation of hepatitis A virus, poliovirus and coliphage MS2 by lime softening and chlorine/monochloramine disinfection. In: Proc. 1993 Water Quality Technology Conference, pp. 479-491. Amer. Water Works Assoc., Denver, CO.

Begosian G.; Bourneuf E.V. 2001. A matter of bacterial life and death. *EMBO Reports* 2:770-774.

Begosian G.; Aardema N.D.; Bourneuf E.V.; Morris P.J.L; O'Neil J.P. 2000. Recovery of hydrogen peroxide-sensitive culturable cells of *Vibrio vulnificus* gives the appearance of resuscitation from a viable but nonculturable state. *J. Bacteriol.* 182:5070-5075.

Bej A.K.; Harold N.; Vickery M.C.L.; Brasher C.; Jeffreys A.; Jones D.D; Depaola A.; Cook D.W. 1997. Use of PCR to determine genomic diversity and distribution of siderophore-mediated iron acquisition genes in clinical and environmental isolates of *Vibrio vulnificus*. *Abstr. Annu. Meet. Amer. Soc. Microbiol.* Q177, p. 485.

Berman D.; Rice E.W.; Hoff J.C. 1988. Inactivation of particle-associated coliforms by chlorine and monochloramine. *Appl. Environ. Microbiol.*, 54(2), 507-512.

Beumer R.R.; de Vries J.; Rombouts F.M. 1992. *Campylobacter jejuni* non-culturable coccoid cells. *Intern. J. Food Microbiol.* 15:153-163.

Brauns L.; Oliver J.D. 1994. Polymerase chain reaction detection of whole cell lysates of *Vibrio vulnificus*. 1994. *Food Biotechnol.* 8:1-6.

Brauns L.; Hudson M.; Oliver J.D. 1991. Use of the polymerase chain reaction in the detection of culturable and nonculturable cells of *Vibrio vulnificus*. *Appl. Environ. Microbiol.* 57:2651-2655.

Breeuwer P.; Abee T. 2000. Assessment of viability of microorganisms employing fluorescence techniques. *Intern. J. Food Microbiol.* 55:193-200.

Bruus J.H.; Nielsen P.H.; Keiding K. 1992. On the stability of activated sludge flocs with implications to dewatering. *Wat. Res.* 26(12), 1597-1604.

Bukhari Z.; Hargy T.M.; Bolton J.R.; Dussert B.; Clancy J.L. 1999. Medium-pressure UV light for oocyst inactivation. *J AWWA* 91(3): 86-94.

Campbell A.T.; Robertson L.J.; Snowball M.R.; Smith H.V. 1995. Inactivation of oocysts of *Cryptosporidium parvum* by ultraviolet light. *Wat. Res.* 29: 2583-2586.

Camper A.K.; McFeters G.A. 1979. Chlorine injury and the enumeration of waterborne coliform bacteria. *Appl. Environ. Microbiol.* 37:633-641.

Camper A.K.; LeChevallierr M.W.; Broadaway S.C.; McFeters G.A. 1985. Evaluation of procedures to desorb bacteria from granular activated carbon. *J. Microbiol. Methods,* 3, 187-198.

Camper A.K.; LeChevallierr M.W.; Broadaway S.C.; McFeters G.A. 1986. Bacteria associated with granular activated carbon particles in drinking water. *Appl. Environ. Microbiol.*, **52**(3), 434-438.

Chang S.L.; Berg; Clarke N.A.; Kabler P.W. 1960. Survival and protection against chlorination of human enteric pathogens in free living nematodes isolated from water supplies. *Amer. J. Trop. Med. Hyg.*, 9:366-371.

Chang J.C.H.; Osoff S.F.; Lobe D.C.; Dorfman M.H.; Dumais C.M.; Qualls R.G.; Johnson J.D. 1985. UV inactivation of pathogenic and indicator microorganisms. *Appl. Environ. Microbiol.* 49(6): 1361.

Chowdhury Z.K.; VanGelder A.; Lawler D.; Moran M. 2000. *Particle Count Method Development for Concentration Standards and Sample Stabilization.* AWWA Research Foundation and American Water Works Association, Denver, CO.

Clancy J.L.; Bukhari Z.; Hargy T.M.; Bolton J.; Dussert B.; Marshall M.M. 2000. Using UV to inactivate Cryptosporidium. *J AWWA* 92(9): 97-104.

Clark R.M.; Hurst C.J.; Regli S. 1993. Costs and benefits of pathogen control in drinking water. p. 181-198. In Craun, G. F. (ed.), *Safety of Water Disinfection: Balancing Chemical and Microbial Risks*. International Life Sciences Press, Washington, D.C.

Coleman S.; Oliver J.D. 1996. Optimization of conditions for the polymerase chain reaction amplification of DNA from culturable and nonculturable cells of *Vibrio vulnificus*. *FEMS Microbiol. Ecol.* 19:127-132.

Colwell, R.R.; Brayton P.R.; Grimes D.J.; Roszak D.R.; Huq S.A.; Palmer L.M. 1985. Viable, but non-culturable *Vibrio cholerae* and related pathogens in the environment: implication for release of genetically engineered microorganisms. *Bio/Technology* 3:817-820.

Colwell R.R.; Brayton P.R.; Herrington D.; Tall B.D.; Huq A.; Levine M.M. 1996. Viable but nonculturable *Vibrio cholerae* O1 revert to a culturable state in human intestine. *World J. Microb. Biotechnol.* 12:28-31.

Créach V.; Baudoux A.C.; Bertru G.; Le Rouzic B. 2003. Direct estimate of active bacteria: CTC use and limitations. *J. Microbiol. Meth.* 52:19-28.

Dale N.G. 1974. Bacteria in intertidal sediments: Factors related to their distribution. *Limnol. Oceanog.* 19, 509-518.

Das G.; Kaveri S.; Das J. 1981. Repair of ultraviolet-light-induced DNA damage in *Vibrio cholerae*. *Biochimica et Biophysica Acta* 655: 413-420.

Day A.; Oliver J.D. 1999. Changes in membrane fatty acid composition of *Vibrio vulnificus* during entry into the VBNC state and following osmotic stress. *Annu. Meet. Amer. Soc. Microbiol.*

Desmonts C.; Minet J., Colwell R.; Cormier M. 1990. Fluorescent-antibody method useful for detecting viable but nonculturable *Salmonella* spp. in chlorinated wastewater. *Appl. Environ. Microbiol.* 56:1448-1452.

Emerick R.W.; Loge F.J.; Thompson D.; Darby J.L. 1999. Factors Influencing Ultraviolet Disinfection Performance Part II. Association of Coliform Bacteria with Wastewater Particles. *Wat. Environ. Res.*, 71(6)1178.

Eriksson L.; Alm B. 1991. Study of flocculation mechanisms by observing effects of a complex agent on activated sludge properties. *Wat. Sci. Tech.* 24(7), 21-28.

Faegri A.; Torsvik V.L.; Goksoyr F. 1977. Bacterial and fungal activities in soil: Separation of bacteria and fungi by a rapid fractionated centrifugation technique. *Soil Biol. Biochem.,* **9**, 105-112.

Federighi M.; Tholozan J.L.; Cappelier J.M.; Tissier J.P.; Jouve J.L. 1998. Evidence of non-coccoid viable but non-culturable *Campylobacter jejuni* cells in microcosm water by direct viable count, CTC-DAPI double staining, and scanning electron microscopy. *Food Microbiol.* 15:539-550.

Finch G.R.; Black E.K.; Gyurek L.L. 1995. Ozone and chlorine inactivation of Cryptosporidium. In *Proc. 1994 Water Quality Technology Conf.,* pp. 1301-1318. Amer. Water Works Assoc., Denver, CO.

Friedberg E.C.; Walker G.C.; Siede W. 1995. *DNA repair and mutagenesis*. ASM press. Washington, D.C.

Fujioka R.S.; Tenno K.M.; Loh P.C. 1983. Mechanism of chloramine inactivation of poliovirus: a concern for regulators? pp. 1067-1076. In: *Water Chlorination: Environmental Impact and Health Effects*, Vol. 4, book 2. Jolley R.L.; Brungs W.A.; Cotruvo J.A.; Cumming R.B.; Mattice J.S.; Jacobs V.A. (eds.). Ann Arbor Science Publishers, Ann Arbor, MI.

Furuse K.; Sakarai T.; Watanabe I. 1967. In: Wang S.Y. (ed.) *Photochemistry and Photobiology of Nucleic Acids*. Academic Press, New York.

Gates F.L. 1929. A Study of the Bactericidal Action of Ultra Violet Light, I. The Reaction to Monochromatic Radiation. *J. Gen. Physiol.* 13:231.

Gibson III C.J.; Hass C.N.; Rose J.B. 1998. Risk assessment of waterborne protozoa: current status and future trends. *Parasitol.* 117 Suppl: S205-212.

Glockner F.O.; Amann R.; Alfreider A.; Pernthaler J.; Psenner R.; Trebesius K.; Schleifer K.H. 1996. An in situ hybridization protocol for detection and identification of planktonic bacteria. System. *Appl. Microbiol.* 19, 403-406.

Graves S.; Bej A.K.. 1994. Use of polymerase chain reaction (PCR) in distinguishing live *Salmonella typhimurium* from biocide-treated dead cells in water. *Abstr. Annu. Meet. Amer. Soc. Microbiol.* Q-259, p. 434.

Gupte A.R.; Joseph S.W. 2001. Evidence for viable non-culturable *Salmonella enterica* serotype Typhimurium DT 104 and its eventual resuscitation. Q-117, p. 606. *Abstr. Annu. Meet. Amer. Soc. Microbiol.*

Harm W. 1980. *Biological Effects of Ultraviolet Radiation*. Cambridge University Press, New York.

Harris G.D.; Adams V.D.; Sorensen D.L.; Curtis M.S. 1987. Ultraviolet inactivation of selected viruses with photoreactivation of the bacteria. *Wat. Res.* 21: 687-692.

Havelaar A.H.; Meulemans C.C.E.; Pot Hogeboom W.M.; Koster J. 1990. Inactivation of bacteriophage MS2 in wastewater effluent with monochromatic and polychromatic UV light. *Wat. Res.* 24: 1387-1394.

Havelaar A.H.; Nieuwstad T.J.; Meulemans C.C.E.; Van Olphen M. 1991. F-specific RNA bacteriophages as model viruses in UV disinfection of wastewater. *Wat. Sci. Technol.* 24: 347-352.

Hoyer O. 1998. Testing performance and monitoring of UV systems for drinking water disinfection. *Wat. Suppl.* 16(1-2): 424-429.

Jacangelo J.G.; Olivieri V.P. 1985. Aspects of the mode of action of monochloramine, pp. 575-586. In: W*ater Chlorination: Chemistry, Environmental Impact and Health Effects*, vol. 5. Jolley R.L., R.J. Bull, W.P. Davis, S. Katz, M.H. Roberts, Jr. and V.A Jacobs (eds.). Lewis Publishers. Chlesea, MI.

Jagger J.H. 1967. *Introduction to Research in UV Photobiology*, Prentice-Hall, Inc., Englewood Cliffs, N.J.

Kell D.B.; Kapreylants A.S.; Weichart D.H.; Harwood C.L; Barer M.R. 1998. Viability and activity in readily culturable bacteria: a review and discussion of the practical issues. *Ant. van Leeuvenhoek* 73:169-187.

King C.H.; Schotts E.B.; Woodley R.E.; Porter K.G. 1988. Survival of coliforms and bacteria pathogens within protoza during chlorination. *Appl. Environ. Microbiol.*, 54:3023-3033.

Kogure K.; Simidu U.; Taga N. 1979. A tentative direct microscopic method for counting living marine bacteria. *Can. J. Microbiol.* 25:415-420.

Kong I.S., Huelsmann A.; Bates T.C.; Hassan H.; Oliver J.D. The role of reactive oxygen species in the viable but nonculturable state in *Vibrio vulnificus*. Submitted to *Appl. Environ. Microbiol.* 2004.

Korich D.G.; Mead J.R.; Madore M.S.; Sinclair N.A.; Sterling C.R. 1990. Effects of ozone, chlorine dioxide, chlorine, and monochloramine on Cryptosporidium oocyst viability. *Appl. Environ. Microbiol*. 56(5): 1423-1428.

LeChevallier M.W.; Norton W.D. 1995. *Giardia* and *Cryptosporidium* in raw and finished water. *J AWWA* 87 (9): 54-68.

LeChevallier M.W.; Evans T.M.; Ramon J.S. 1981. Effect of turbidity on chlorination efficiency and bacterial persistence in drinking water. *Appl. Environ. Micrbiol*., 42(1), 159-167.

LeChevallier M.W.; Hassenauer T.S.; Camper A.K.; McFeters G.A. 1984. Disinfection of bacteria attached to granular activated carbon. *Appl. Environ. Microbiol.,* 48(5), 918-923.

LeChevallier M.W.; Singh A.; Shciemann D.A.; McFeters G.A. 1985. Changes in virulence of water enteropathogens with chlorine injury. *Appl. Environ. Microbiol.* 50:412-219.

Linden K.G.; Darby J.L. 1997. Estimating the effective germicidal dose from medium pressure UV lamps. *ASCE: J. Environ. Eng.* 123(11)1142

Linden K.G.; Darby J.L. 1998. UV disinfection of marginal effluents: determining UV absorbance and subsequent estimation of UV intensity. *Wat. Environ. Res.* 70(2)214.

Linden K.G.; Shin G.A.; Faubert G.; Cairns W.; Sobsey M.D. 2002. UV disinfection of *Giardia lamblia* in water. *Environ. Sci. Technol.*, 36(11)2519.

Lindenauer K.G.; Darby J.L. 1994. Ultraviolet disinfection of wastewater: Effect of dose on subsequent photoreactivation. *Wat. Res.*, 28:805-817.

Lleò M.M.; Bonato B.; Tafi M.C.; Signoretto C.; Boaretti M.; Canepari P. 2001. Resuscitation rate in different enterococcal species in the viable but non-culturable state. *J. Appl. Microbiol.* 91:1095-1102.

Lleò M.M.; Pierobon S.; Tafi M.C.; Signoreto, C.; Canepari P. 2000. mRNA detection by reverse transcription-PCR for monitoring viability over time in an *Enterococcus faecalis* viable but nonculturable population maintained in a laboratory microcosm. *Appl. Environ. Microbiol.* 66:4564-4567.

Loge F.; Emerick R.; Williams C.; Kido W.; Darby J. 1997. Impact of particle associated coliform on UV disinfection. *Proc. WEF Annual Conference*, Chicago, IL, CP 1701, WEF, Alexandria, VA.

Mackey B.M. 2000. Injured bacteria. p. 315-341, In: *The microbiological safety and quality of food.* Lund, B.M., A. Baird-Parker, and G.M. Gould, ed. Aspen Publ. Inc., Gaithersburg, MD.

McDougald D.; Prosser J.I.; Glover L.A.; Oliver J.D. 1995. Effect of temperature and plasmid carriage on nonculturability in organisms targeted for release. *FEMS Microbiol. Ecol.* 17:229-238.

McFeters G.A. 1990. Enumeration, occurrence, and significance of injured indicator bacteria in drinking water. pp. 478-492 In: *Drinking Water Microbiology: Progress and Recent Developments.* McFeters G.A. (ed.). Spriner-Verlag, N.Y.

McFeters G.A.; LeChevalier M.W. 2000. Chemical disinfection and injury of bacteria in water. pp. 255-275, In: Colwell R.R.; Grimes D.J., eds., *Nonculturable Microorganisms in the Environment*. Amer. Soc. Microbiol. Press, Washington, D.C.

McFeters G.A.; Yu F.P.; Pyle B.H.; Stewart P.S. 1995. Physiological assessment of bacteria using fluorochromes. *J. Microbiol. Meth.* 21:1-13.

McGovern V.P.; Oliver J.D. 1995. Induction of cold-responsive proteins in *Vibrio vulnificus*. *J. Bacteriol.* 177:4131-4133.

Melnick J.L.; Gerba C.P. 1980. The ecology of enteroviruses in natural water. CRC. *Critical Reviews in Environmental Control* 10 (1): 65-93.

Meng Q.S.; Gerba C.P. 1996. Comparative inactivation of enteric adenovirus, poliovirus and coliphages by ultraviolet radiation. *Wat. Res.* 30: 2665-2668.

Mizunoe Y.; Wai S.N.; Ishikawa T.; Takade A.; Yoshida S.-i. 2000. Resuscitation of viable but nonculturable cells of *Vibrio parahaemolyticus* induced at low temperature under starvation. *FEMS Microbiol. Lett.* 186:115-120.

Mizunoe Y.; Wai S.N.; Takade A.; Yoshida S.-i., 1999. Restoration of culturability of starvation-stressed and low-temperature-stressed *Escherichia coli* O157 cells by using H_2O_2-degrading compounds. *Arch. Microbiol.* 172:63-67.

Morowitz H.J. 1950. Absorption effects in volume irradiation of microorganism. *Science* 111: 229.

Morton D.; Oliver J.D., 1994. Induction of carbon starvation proteins in *Vibrio vulnificus*. *Appl. Environ. Microbiol.* 60:3653-3659.

Mukamolova G.V.; Yanopolskaya N.D.; Kell D.B.; Kaprelayants A.S. 1998a. On resuscitation from the dormant state of *Micrococcus luteus*. *Ant. Van Leeuwen.* 73:237-243.

Mukamolova G.V.; Kaprelayants A.S.; Young D.I.; Young M.; Kell D.B. 1998b. A bacterial cytokine. *Proc. Natl. Acad. Sci.* USA 95:8916-8921.

Mukamolova G.V.; Kormer S.S.; Kell D.B.; Kaprelayants A.S. 1999. Stimulation of the multiplication of *Micrococcus luteus* by an autocrine growth factor. *Arch Microbiol.* 172:9-14.

Nieuwstad T.J.; Havelaar A.H. 1994. The kinetics of batch ultraviolet inactivation of bacteriophage MS2 and microbiological calibration of an ultraviolet pilot plant. *J Environ. Sci. Heal.* 29:257-273.

Nieuwstad T.J.; Havelaar A.H.; Van Olphen M. 1991. Hydraulic and microbiological characterization of reactors for UV disinfection of secondary effluent. *Wat. Res.* 25:775-784.

Nwoguh C.E.; Harwood C.R.; Barer M.R. 1995. Detection of induced β-galactosidase activity in individual nonculturable cells in pathogenic bacteria by quantitative cytological assay. *Molec. Micrbiol.* 17:545-554.

O'Brien R.T.; Newman J. 1979. Structural and compositional changes associated with chlorine inactivation of poliovirus. *Appl. Environ. Microbiol.*, 38:1034-1039.

Oguma K.; Katayama H.; Mitani H.; Morita S.; Hirata T.; Ohgaki S. 2001. Determination of pyrimidine dimers in *Escherichia coli* and *Cryptosporidium parvum* during UV light inactivation, photoreactivation and dark repair. *Appl. Environ. Microbiol.*, **67**, 4630.

Oguma K.; Katayama H.; Ohgaki S. 2002. Photoreactivation of *Escherichia coli* after low- or medium-pressure UV disinfection determined by an endonuclease sensitive site assay *Appl. Environ. Microbiol.*, 68(12)6029-6035

Oliver B.G.; Cosgrove E.G. 1975.The disinfection of sewage treatment plant effluents using ultraviolet light. *Can J. Chem. Eng.*, **53**, 170.

Oliver J.D.; Wanucha D. 1989. Survival of *Vibrio vulnificus* at reduced temperatures and elevated nutrient. *J. Food Safety* 10:79-86.

Oliver J.D. 1993. Formation of viable but nonculturable cells. pp. 239-272 In: *Starvation in Bacteria*, Kjelleberg S. (ed.). Plenum Press.

Oliver J.D. 1995. The viable but nonculturable state in the human pathogen, *Vibrio vulnificus. FEMS Lett.* (Minireview). 133:203-208.

Oliver J.D. 2000a. Public health significance of viable but nonculturable bacteria. In: *Non-Culturable Microorganisms in the Environment*, Colwell R.R.; Grimes D.J. (ed.). Chapman and Hall, Publ.

Oliver J.D. 2000b. Problems in detecting dormant (VBNC) cells, and the role of DNA elements in this response. In: *Tracking Genetically Engineered Microorganisms*. Jansson J.K.; van Elsas J.D.; Bailey M. (ed). Landes Bioscience, Georgetown, TX

Oliver, J.D. 2000c. The viable but nonculturable state and cellular resuscitation. pp. 723-730, In: Bell C.R.; Brylinsky M.; Johnson-Green P. eds., *Microbial Biosystems: New Frontiers*. Atlantic Canada Soc. Microb. Ecol., Halifax, Canada.

Oliver J.D.; Bockian R. 1995. In vivo resuscitation, and virulence towards mice, of viable but nonculturable cells of *Vibrio vulnificus*. *Appl. Environ. Microbiol.* 61:2620-2623.

Oliver J.D.; Hite F.; McDougald D.; Andon N.L.; Simpson L.M. 1995. Entry into, and resuscitation from, the viable but nonculturable state by *Vibrio vulnificus* in an estuarine environment. *Appl. Environ. Microbiol.* 61:2624-2630.

Oliver J.D.; Nilsson L.; Kjelleberg S. 1991. The formation of nonculturable cells of *Vibrio vulnificus* and its relationship to the starvation state. *Appl. Environ. Microbiol.* 57:2640-2644.

Olivieri V.P.; Kruse C.W.; Hsu Y.-C.; Griffiths A.C.; Kawata K. 1975. The comparative mode of action of chlorine, bromine and iodine on f2 bacterial virus, pp. 145-162. In: Johnson J.D. (ed.) *Disinfection: Water and Wastewater*. Ann Arbor Science Publishers, Ann Arbor.

Olson B.H. Stuart M.H. 1990. Factors that change bacterial resistance to disinfection, pp. 885-904. In: *Water Chlorination: Chemistry, Environmental Impact and health Effects*, vol. 6. Jolly R.L.; Condie L.W. ; Johnson J.D.; Katz S.; Minear R.A.; Mattice J.S.; Jacobs V.A. (eds.) Lewis Publishers, Chelsea, Mi.

Parker J.A.; Darby J.L. 1995. Particle-associated coliform in secondary effluents: shielding from ultraviolet light disinfection. *Wat. Env. Res.*, **67**(7), 1065-1075.

Pommepuy M.; Butin M.; Derrien A.; Gourmelon M.; Colwell R.R.; Cormier M. 1996. Retention of enteropathogenicity by viable but nonculturable *Escherichia coli* exposed to seawater and sunlight. *Appl. Environ. Microbiol.* 62:4621-4626.

Porter J.; Edwards C.; Pickup R.W. 1995. Rapid assessment of physiologicalstatus in *Escherichia coli* using fluorescent probes. *J. Appl. Bacteriol.* 4:399-408.

Qualls R.G.; Ossoff S.F.; Chang J.C.H.; Dorfman M.H.; Dumais C.M.; Lobe D.C.; Johnson J.D. 1985. Factors controlling sensitivity in ultraviolet disinfection of secondary effluents. *J. Wat. Pollut. Control Fed.*, 57:1006-1011.

Rahman I.; Shahamat M.; Kirchman P.A.; Russek-Cohen E.; Colwell R.R. 1994. Methionine uptake and cytopathogenicity of viable but nonculturable *Shigella dysenteriae* Type 1. *Appl. Environ. Microbiol.* 60:3573-3578.

Ramsay A.J. 1984. Extraction of bacteria from soil: Efficiency of shaking or ultrasonication as indicated by direct counts and autoradiography. *Soil. Biol. Biochem.*, **16**(5), 475-481.

Ransome M.E.; Whitmore T.N.; Carrington E.G. 1993. Effect of disinfectants on the viability of *Cryptosporidium parvum*. *Wat. Suppl.* 11: 103-117.

Regnault B.; Martin-Delautre S.; Grimont P.A.D. 2000. Problems associated with the direct viable count procedure applied to gram-positive bacteria. Intern. *J. Food Microbiol.* 55:281-184.

Saux M.F.-L.; Hervio-Heath D.; Loaec S.; Colwell R.R.; Pommepuy M. 2002. Detection of cytotoxin-hemolysin mRNA in nonculturable populations of environmental and clinical *Vibrio vulnificus* strains in artificial seawater. *Appl. Environ. Microbiol.* 68:5641-5646.

Shin G.; Linden K.G.; Arrowood M.J.; Sobsey M.D. 2001. Low pressure UV inactivation and DNA repair potential of Cryptosporidium parvum oocysts. *Appl. Environ. Microbiol.* 67:3029-3032.

Sides L.; Hite F.; Oliver J.D. 1999. Effects of reactive oxygen species (ROS) inhibitors on resuscitation of viable but nonculturable (VBNC) cells of *Vibrio vulnificus, Escherichia coli* O157:H7 and non-O157:H7. Q129, p. 558, *Annu. Meet. Amer. Soc. Microbiol.*

Signoretto C.; Lleò M.M.; Tafi M.C.; Canepari P. 2000. Cell wall chemical composition of *Enterococcus faecalis* in the viable but nonculturable state. *Appl. Environ. Microbiol.* 66:1953-1959.

Signoretto C.; Lleò M.M.; Canepari P. 2002. Modification of the peptidoglycan of *Escherichia coli* in the viable but nonculturable state. *Curr. Microbiol.* 44:125-131.

Singh A.; McFeters G.A. 1990. Injury of enteropathogenic bacteria in drinking water. pp. 368-379 In: *Drinking Water Microbiology*. Progress and Recent Developments. McFeters G.A. (ed.). Spriner-Verlag, NY.

Singh A.; Yeager R.; McFeters G.A. 1986. Assessment of in vivo revival, growth, and pathogenicity of *Escherichia coli* strains after copper- and chlorine-induced injury. *Appl. Environ. Microbiol.* 52:832-837.

Sobsey M.D.; Carrick R.J.; Jensen H.R. 1978. Improved methods for detecting enteric viruses in oysters. *Appl. Environ. Microbiol.* 36(1):121-128

Sobsey M.D. 1989. Inactivation of Health-related Microorganisms in Water by Disinfection Processes. *Wat. Sci. Technol.* 21:179-195.

Sobsey M.D.; Fuji T.; Hall R.M. 1991. Inactivation of cell-associated and dispersed hepatitis A virus in water. *J. AWWA*, 83(11):64-67.

Sommer R.; Haider T.; Cabaj A.; Heidenreich E.; Kundi M. 1996. Increased inactivation of *Sacharomyces cerevisia*e by protraction of UV irradiation. *Appl. Environ. Microbiol.*, **62**, 1977.

Sommer R.; Haider T.; Cabaj A.; Pribal W.; Lhotsky M. 1998. Time dose reciprocity in UV disinfection of water. *Wat. Sci. Tech.*, **38**(12), 145.

Steinert M.; Emody L.; Amann R.; Hacker J. 1997. Resuscitation of viable but nonculturable *Legionella pneumophila* Philadelphia JR32 by *Acanthamoeba castellanii. Appl. Environ. Microbiol.* 63:2047-2053.

Stevenson I.L. 1958. The effect of sonic vibration on the bacterial plate count of soil. *Plant and Soil*, 10:1-8.

Stewart M.H. 1983. Plasmid-mediated hydrocarbon utilization in Acinetobacter phosphadevorus, Masters Thesis, University of Arizona, Tucson.

Stewart M.H. 1991 Physiological mechanisms of bacterial resistance to inorganic monochloramine, Ph.D. Dissertation, 234 pages, University of California-Irvine, Irvine

Stewart M.H.; Olson B.H. 1992. Physiological studies of chloramine resistance developed by *Klebsiella pneumoniae* under low nutrient conditions. *Appl. Environ. Microbiol.,* 58:2918-2927.

Stewart M.H.; Olson B.H. 1996. Bacterial resistance to potable water disinfectants, pp. 140-192. In: *Modeling Disease Transmission and Its Prevention by Disinfection*. Hurst C.J. (ed.). Cambridge University Press, Cambridge.

Tenno K.M.; Fujioka R.S.; Loh P.C. 1979. The mechanism of poliovirus inactivation by hyprchlorous acid, p. 665-675. In: R. Jolley, Brungs W.A., and Cumming R.B. (ed.) *Proceedings of the Third Conference on Water Chlorination: environmental impact and health effects*. Ann Arbor Science Publishers, Inc., Ann Arbor, MI.

Tholozan J.L.; Cappelier J.M.; Tissier J.P.; Delattre G.; Federighi M. 1999. Physiological characterization of viable-but-nonculturable *Camplyobacter jejuni* cells. *Appl. Environ. Microbiol.* 65:1110-1116.

Tracy H.W.; Camarena V.M.; Weing F. 1966. Coliform persistence in highly chlorinated waters. *J. AWWA*, 58:1151-1159.

Tree J.A.; Adams M.R.; Lees D.L. 1997. Virus inactivation during disinfection of wastewater by chlorination and UV irradiation and the efficacy of F+ bacteriophages as a 'viral indicator'. *Wat. Sci. Technol.* 35(11-12):227-232.

U.S. EPA. 1986. Design Manual: municipal wastewater disinfection. EPA/625/1-81/021. U. S. Environmental Protection Agency, Cincinnati, OH.

Vedeikis E.; Oliver J.D. 2000. The role of reactive oxygen species (ROS) in the cultivation of *Vibrio vulnificus* present in the viable but nonculturable state. *Ann. Meet. Amer. Soc. Microbiol.*

Venczel L.V.; Sobsey M.D.; Crawford-Brown D. 1992. The inactivation kinetics of monochloramine on monodispersed hepatitis A virus and MS2. In: *Proc. 1991 Water Quality Technology Conference*, pp. 531-550. Amer. Water Works Assoc., Denver, CO.

Wai S.N.; Moriya T.; Kondo K.; Misumi H.; Amako K. 1996. Resuscitation of *Vibrio cholerae* O1 strain TSI-4 from a viable but nonculturable state by heat shock. *FEMS Microbiol. Lett.* 136:187-191.

Warner J.M.; Oliver J.D. 1998. Randomly amplified polymorphic DNA analysis of starved and viable but nonculturable *Vibrio vulnificus* cells. *Appl. Environ. Microbiol.* 64:3025-3028.

Water Environment Federation, 1996. *Wastewater Disinfection: Manual of Practice FD-10.* WEF, Alexandria, VA.

Weise W.; Rheinheimer G. 1979. Fluoreszenz-mikroskopische untersuchungen uber die bakterien-besiedlung mariner sandsedimente. *Botanica Marina* **22**, 99-106.

Whitby G.E.; Palmateer G.; Cook W.G.; Maarschalkerweerd J.; Huber D.; Flood K. 1984. Ultraviolet disinfection of secondary effluent. *J. Wat. Pollut. Contr. Fed.* 56:844-850.

Whitesides M.D.; Oliver J.D. 1997. Resuscitation of *Vibrio vulnificus* from the viable but nonculturable state. *Appl. Environ. Microbiol*. 64:3025-3028.

Wiedenmann A.; Braun M.; Botzenhart K. 1997. Evaluation of the disinfection potential of low chlorine concentration in tap water using immobilized *Enterococcus faecium* in a continuous flow device. *Wat. Sci. Tech.,* **35**(11-12), 77.

Wilson B.R.; Roessler P.F.; Van Dellen E.; Abbaszadegan M.; Gerba C.P. 1992. Coliphage MS-2 as a UV disinfection efficacy test surrogate for bacteria and viral pathogens. p 219-235. In *Proceedings of Water Quality Technology Conference*. American Water Works Association. Denver, CO.

Winkler U. 1964. Host-cell reactivation of lethal damage induced by ultraviolet light and X-rays in *Serratia marcescens* Hy and its phage kappa. *Virol*. 24:518-520

Wolf P.; Oliver J.D. 1992. Temperature effects on the viable but nonculturable state of *Vibrio vulnificus*. *FEMS Microbiol. Ecol.* 101:33-39.

Yaron S.; Matthews K. 2002. A reverse transcriptase-polymerase chain reaction assay for detection of viable *Escherichia coli* O157:H7: investigation of specific target genes. *J. Appl. Microbiol.* 92:633-640.

Young D.C.; Sharp D.G. 1985. Virion conformational forms and the complex inactivation kinetics of echovirus by chlorine in water. *Appl. Environ. Microbiol.,* 49:359-365.

Zimmer J.; Slawson R.M. 2002. Potential repair of *Escherichia coli* DNA following exposure to UV radiation from both medium- and low-pressure UV source used in drinking water treatment. *Appl. Environ. Microbiol.* 68(7):3293-3299.

Zimmer J.L.; Slawson R.M.; Huck P.M. 2003. Inactivation and potential repair of Cryptosporidium parvum following low- and medium-pressure ultraviolet irradiation. *Wat. Res.* 37(14): 3517-3523.

Zvyagintsev D.G; Galkina V.E. 1967. Ultrasonic treatment as a method for preparation of soil for microbiological analysis. *Mikrobiologiya***, 36**, 910-916.

WASTEWATER UTILITY

Alabama
Montgomery Water Works & Sanitary Sewer Board

Alaska
Anchorage Water & Wastewater Utility

Arizona
Gila Resources

Glendale, City of, Utilities Department

Mesa, City of

Peoria, City of

Pima County Wastewater Management

Arkansas
Little Rock Wastewater Utility

California
Calaveras County Water District

Central Contra Costa Sanitary District

Contra Costa Water District

Crestline Sanitation District

Delta Diablo Sanitation District

Dublin San Ramon Services District

East Bay Dischargers Authority

East Bay Municipal Utility District

El Dorado Irrigation District

Fairfield-Suisun Sewer District

Fresno Department of Public Utilities

Irvine Ranch Water District

Las Virgenes Municipal Water District

Livermore, City of

Lodi, City of

Los Angeles, City of

Los Angeles County, Sanitation Districts of

Napa Sanitation District

Orange County Sanitation District

Palo Alto, City of

Riverside, City of

Sacramento Regional County Sanitation District

San Diego Metropolitan Wastewater Department, City of

San Francisco, City & County of

San Jose, City of

Santa Barbara, City of

Santa Rosa, City of

South Bayside System Authority

South Coast Water District

South Orange County Wastewater Authority

Sunnyvale, City of

Union Sanitary District

West Valley Sanitation District

Colorado
Boulder, City of

Colorado Springs Utilities

Littleton/Englewood Water Pollution Control Plant

Metro Wastewater Reclamation District, Denver

Connecticut
The Mattabasset District

New Haven, City of, WPCA

District of Columbia
District of Columbia Water & Sewer Authority

Florida
Broward, County of

Fort Lauderdale, City of

Gainesville Regional Utilities

JEA

Miami-Dade Water & Sewer Authority

Orange County Utilities Department

Orlando, City of

Reedy Creek Improvement District

Seminole County Environmental Services

St. Petersburg, City of

Stuart Public Utilities

Tallahassee, City of

Tampa, City of

Toho Water Authority

West Palm Beach, City of

Georgia
Atlanta Department of Watershed Management

Augusta, City of

Clayton County Water Authority

Cobb County Water System

Columbus Water Works

Fulton County

Gwinnett County Department of Public Utilities

Savannah, City of

Hawaii
Honolulu, City and County of

Illinois
American Bottoms Wastewater Treatment Plant

Greater Peoria Sanitary District

Kankakee River Metropolitan Agency

Metropolitan Water Reclamation District of Greater Chicago

North Shore Sanitary District

Wheaton Sanitary District

Iowa
Ames, City of

Cedar Rapids Wastewater Facility

Des Moines Metro Wastewater Reclamation Authority

Iowa City, City of

Kansas
Johnson County Unified Wastewater Districts

Unified Government of Wyandotte County/Kansas City, City of

Kentucky
Louisville & Jefferson County Metropolitan Sewer District

Louisiana
Sewerage & Water Board of New Orleans

Maine
Bangor, City of

Portland Water District

Maryland
Anne Arundel County Bureau of Utility Operations

Howard County Department of Public Works

Washington Suburban Sanitary Commission

Massachusetts
Boston Water & Sewer Commission

Upper Blackstone Water Pollution Abatement District

Michigan
Ann Arbor, City of

Detroit, City of

Holland Board of Public Works

Lansing, City of

Owosso Mid-Shiawassee County WWTP

Saginaw, City of

Wayne County Department of Environment

Wyoming, City of

Minnesota
Rochester, City of

Western Lake Superior Sanitary District

Missouri
Independence, City of

Kansas City Missouri Water Services Department

Little Blue Valley Sewer District

Nebraska
Lincoln Wastewater System

Nevada
Henderson, City of

New Jersey
Bergen County Utilities Authority

Passaic Valley Sewerage Commissioners

New York
New York City Department of Environmental Protection

Rockland County Solid Waste Management Authority/ Sewer District

North Carolina
Charlotte/Mecklenburg Utilities

Durham, City of

Metropolitan Sewerage District of Buncombe County

Orange Water & Sewer Authority

Ohio
Akron, City of

Butler County Department of Environmental Services

Columbus, City of

Metropolitan Sewer District of Greater Cincinnati

Northeast Ohio Regional Sewer District

Summit, County of

Oklahoma
Tulsa, City of

Oregon
Clean Water Services

Eugene/Springfield Water Pollution Control

Water Environment Services

Pennsylvania
Philadelphia, City of

University Area Joint Authority, State College

South Carolina
Charleston Commissioners of Public Works

Mount Pleasant Waterworks & Sewer Commission

Spartanburg Sanitary Sewer District

Tennessee
Cleveland, City of

Knoxville Utilities Board

Murfreesboro Water & Sewer Department

Nashville Metro Water Services

Texas
Austin, City of

Dallas Water Utilities

Denton, City of

El Paso Water Utilities

Fort Worth, City of

Gulf Coast Waste Disposal Authority

Houston, City of

San Antonio Water System

Trinity River Authority

WERF SUBSCRIBERS

Utah
Salt Lake City Corporation

Virginia
Alexandria Sanitation Authority
Arlington, County of
Fairfax County Virginia
Hampton Roads Sanitation District
Henrico, County of
Hopewell Regional Wastewater Treatment Facility
Loudoun County Sanitation Authority
Lynchburg Regional WWTP
Prince William County Service Authority
Richmond, City of
Rivanna Water & Sewer Authority

Washington
Edmonds, City of
Everett, City of
King County Department of Natural Resources
Seattle Public Utilities
Sunnyside, Port of
Yakima, City of

Wisconsin
Green Bay Metro Sewerage District
Madison Metropolitan Sewerage District
Milwaukee Metropolitan Sewerage District
Racine, City of
Sheboygan Regional Wastewater Treatment
Wausau Water Works

Australia
South Australian Water Corp.
Sydney Water Corp.
Water Corp. of Western Australia

Canada
Greater Vancouver Regional District
Toronto, City of, Ontario
Winnipeg, City of, Manitoba

Mexico
Servicios de Agua y Drenaje de Monterrey, I.P.D.

New Zealand
Watercare Services Limited

Singapore
Singapore Public Utilities Board

United Kingdom
Yorkshire Water Services Limited

STORMWATER UTILITY

California
Los Angeles, City of, Department of Public Works
Monterey, City of
San Francisco, City & County of
Santa Rosa, City of
Sunnyvale, City of

Colorado
Boulder, City of

Georgia
Griffin, City of

Iowa
Cedar Rapids Wastewater Facility
Des Moines Metro Wastewater Reclamation Authority

Kansas
Overland Park, City of

Kentucky
Louisville & Jefferson County Metropolitan Sewer District

Maine
Portland Water District

Minnesota
Western Lake Superior Sanitary District

North Carolina
Charlotte, City of, Stormwater Services

Pennsylvania
Philadelphia, City of

Tennessee
Chattanooga Storm Water Management

Washington
Seattle Public Utilities

STATE

Arkansas Department of Environmental Quality
Fresno Metropolitan Flood Control District, Calif.
Urban Drainage & Flood Control District, Colo.

CORPORATE

ADS Environmental Services
The ADVENT Group Inc.
Alan Plummer & Associates
Alden Research Laboratory
Aqua-Aerobic Systems, Inc.
Aquateam–Norwegian Water Technology Centre A/S
BaySaver Inc.
BioVir Laboratories, Inc.
Black & Veatch
Boyle Engineering Corporation
Brown & Caldwell
Burns & McDonnell
CABE Associates Inc.
The Cadmus Group
Camp Dresser & McKee Inc.
Carollo Engineers Inc.
Carpenter Environmental Associates Inc.
CDS Technologies Inc.
Chemtrac Systems Inc.
CH2M HILL
Damon S. Williams Associates, LLC
David L. Sheridan, P.C.
Dewling Associates, Inc.
Earth Tech Inc.
Eco-Matrix
Ecovation
Environmental Engineers International
EMA Inc.
The Eshelman Company Inc.
Finkbeiner, Pettis, & Strout Inc.
Freese & Nichols, Inc.
ftn Associates Inc.
Fuss & O'Neill, Inc.
Gannett Fleming Inc.
Golder Associates Inc.
Greeley and Hansen LLC
Hazen & Sawyer, P.C.
HDR Engineering Inc.
HNTB Corporation
HydroQual Inc.
Infilco Degremont Inc.
Ingersoll-Rand Energy Systems
Insituform Technologies Inc.
Institute for Environmental Technology & Industry, Korea
Jacobson Helgoth Consultants Inc.
Jason Consultants Inc.
Jordan, Jones, & Goulding Inc.
KCI Technologies Inc.
Kelly & Weaver, P.C.
Kennedy/Jenks Consultants
Komline Sanderson Engineering Corporation
Lawler, Matusky & Skelly Engineers, LLP
Limno-Tech Inc.
Lombardo Associates Inc.
Malcolm Pirnie Inc.
McKim & Creed
MEC Analytical Systems Inc.
Metcalf & Eddy Inc.
MWH
New England Organics
Odor & Corrosion Technology Consultants Inc. (OCTC)
Oswald Green, LLC
PA Government Services Inc.
Parametrix Inc.
Parsons
Post, Buckley, Schuh & Jernigan
R&D Engineering/Conestoga Rover & Associates
The RETEC Group
RMC, Inc. (Raines, Melton & Carella)
R.M. Towill Corporation
Ross & Associates Ltd.
Royce Technologies
SAIC Maritime Technical Group
Short Elliott Hendrickson, Inc.
Stantec Consulting Group, Inc.
Stormwater Management, Inc.
Synagro Technologies, Inc.
Tetra Tech Inc.
Trojan Technologies Inc.
URS Corporation
Veolia Water NATC
Wade-Trim Inc.
Weston Solutions Inc.
Woodard & Curran
WRc/D&B, LLC
WWETCO, LLC
Zoeller Pump Company

INDUSTRY

American Electric Power
ChevronTexaco Energy Research & Technology Company
The Coca-Cola Company
Dow Chemical Company
DuPont Company
Eastman Kodak Company
Eli Lilly & Company
Merck & Company Inc.
ONDEO Services
Procter & Gamble Company
PSEG Services Corp.
RWE Thames Water Plc
Severn Trent Services Inc.
United Water Services LLC

Note: List as of 7/1/04

WERF Product Order Form

As a benefit of joining the Water Environment Research Foundation, subscribers are entitled to receive one complimentary copy of all final reports and other products.Additional copies are available at cost (usually $10).To order your complimentary copy of a report, please write "free" in the unit price column.WERF keeps track of all orders.If the charge differs from what is shown here, we will call to confirm the total before processing.

Name Title

Organization

Address

City State Zip Code Country

Phone Fax Email

Stock #	Product	Quantity	Unit Price	Total

Postage & Handling

VA Residents Add 4.5% Sales Tax

Canadian Residents Add 7% GST

TOTAL

Method of Payment:

(All orders must be prepaid.)

❑ Check or Money Order Enclosed

❑ Visa ❑ Mastercard ❑ American Express

Account No. Exp. Date

Signature

Shipping & Handling:

Amount of Order	United States	Canada & Mexico	All Others
Up to but not more than:	Add:	Add:	Add:
$20.00	$5.00*	$8.00	50% of amount
30.00	5.50	8.00	40% of amount
40.00	6.00	8.00	
50.00	6.50	14.00	
60.00	7.00	14.00	
80.00	8.00	14.00	
100.00	10.00	21.00	
150.00	12.50	28.00	
200.00	15.00	35.00	
More than $200.00	Add 20% of order	Add 20% of order	

*minimum amount for all orders

Note: Please make checks payable to the Water Environment Research Foundation.

To Order (Subscribers Only):

Log on to www.werf.org and click on the "Product Catalog."

Phone: (703) 684-2470
Fax: (703) 299-0742.

WERF
Attn: Subscriber Services
635 Slaters Lane
Alexandria, VA 22314-1177

To Order (Non-Subscribers):

Non-subscribers may be able to order WERF publications either through WEF (www.wef.org) or IWAP (www.iwapublishing.com).Visit WERF's website at www.werf.org for details.

IWA Publishing's authorised EU representative for General Product Safety Regulations is Diane D'Arras, 15 rue Duret, 75116 Paris, France, e-mail: safety@iwap.co.uk.

Printed and bound by CPI Group (UK) Ltd, Croydon, CR0 4YY
12/05/2026
02108868-0012